PHYSICAL ORGANIC CHEMISTRY

PHYSICAL ORGANIC CHEMISTRY
THROUGH SOLVED PROBLEMS

Joseph B. Lambert

NORTHWESTERN UNIVERSITY

HOLDEN-DAY, INC.

San Francisco   London   Dusseldorf
Johannesburg   Panama   Singapore   Sydney

Permission for the reproduction of the figures on pages 142, 160, 164, 168 (2), 196, and 198 has been granted by the American Chemical Society.

PHYSICAL ORGANIC CHEMISTRY
THROUGH SOLVED PROBLEMS

Library of Congress Catalog Card Number: 78-060359
ISBN: 0-8162-4921-0

Printed in the United States of America

234567890   80

PREFACE

Chemistry texts for undergraduates have traditionally offered numerous problems but few explanatory solutions. The trend recently has been to correct this deficiency by providing auxiliary answer books, although too often the answers consist of a number, a structure, or a phrase without explanation. Ideally, an answer should be considerably longer than the original statement of the problem. Texts and lectures provide the framework for learning new material, but deeper understanding is gained through actual application via problem solving. Such experience is most profitable when the student can be confident in the validity of the answer and can fully appreciate the reasoning involved in arriving at the answer, as provided by worked problems.

The problems given in this text are concerned with the principles of physical organic chemistry. Any student who has learned the basic physical chemical principles of kinetics and thermodynamics is ready for this material as a second organic course. The principles and methodology of physical organic chemistry in turn can be applied to the problems of mechanistic organic, inorganic, and biochemistry, and to some extent to synthetic organic chemistry. A possible physical organic course outline, with problem references, is given at the end of this section, followed by an alternative reaction mechanisms reference list. This text assumes no knowledge of mechanistic organic chemistry beyond that normally acquired in a basic organic course, namely electrophilic aromatic substitution, nucleophilic aliphatic substitution, elimination, electrophilic addition, nucleophilic addition (hydrolysis, esterification), carbocations, carbanions, free radicals, carbenes, acid/base equilibria, and pericyclic reactions. This text is not intended to be a survey of the subject of reaction mechanisms but to illustrate the principles whereby such studies can be carried out. Consequently, simple arrow-pushing problems have not been emphasized. Spectroscopy has been excluded, since it is one area that has been adequately supplied in recent years with worked problems.

This volume can serve either as a supplement to a standard text on physical organic chemistry/organic reaction mechanisms or as the sole text for student use when the lecture material is developed from several sources. Although the expected primary role of these problems is as homework to enhance the student's understanding of the lecture material, the professor can also use the problems as material for examinations. If these problems are used for examinations, about three questions are recommended per hour of examination time. If these problems are used by the student as homework exercises, about an hour should be spent on each problem, including examination of the original literature.

Most of the questions have been selected from the physical organic literature of the decade 1968-1978. Literature references are given at the end of each problem. Problems of this sort offer an entry into the literature without the trappings of a full research paper, which is normally directed to fellow research workers rather than to students. The author is indebted to Karen L. Permer, who typed the original manuscript, and to Janet H. Goranson, who typed the final production version.

Joseph B. Lambert
Evanston, Illinois
June, 1978

CONTENTS

REFERENCE OUTLINE
TO PROBLEMS

PHYSICAL ORGANIC PRINCIPLES APPROACH

I. EQUILIBRIUM

  A. Configurational Analysis
    1. Classification of stereoisomers, 1-3, 1-4, 1-6
    2. Prochirality, 1-7

  B. Conformational Analysis
    1. Acyclic conformations
      a. Saturated systems, 1-2, 1-6
      b. Unsaturated systems, 1-9, 3-4, 4-4, 4-5, 4-8, 6-5
    2. Cyclic conformations
      a. Six-membered rings, 1-1, 1-3, 1-4, 1-8, 2-1, 3-1, 3-2, 3-3, 7-4, 7-5
      b. Rings of other than six members, 1-5, 7-5
      c. Ring fusion, 1-1, 1-3, 1-4, 2-1
    3. Sources of strain
      a. Factors common to hydrocarbons, 1-1, 1-2, 1-3, 1-4, 1-5, 1-6, 1-8, 2-1, 3-1, 3-2
      b. Electrostatic, 1-6, 3-1, 3-2, 3-3, 3-4
      c. Orbital, 1-9, 3-1, 3-2, 3-3
  C. Organothermochemistry
    1. Enthalpy vs. entropy, 2-1, 2-2, 4-7

REACTION MECHANISMS APPROACH

1. **Acid/base equilibria,** 1-9, 3-5, 3-6, 3-8, 4-1, 4-2, 4-3, 4-4, 4-6, 4-7, 4-8, 5-1, 5-2, 5-3, 5-4, 5-5, 6-1, 6-2, 7-1, 7-2, 7-5, 9-7, 10-1, 10-2, 10-3, 10-4, 10-5, 10-6, 10-7, 10-8, 10-9, 10-10

2. **Conformational equilibria,** 1-1, 1-2, 1-3, 1-4, 1-5, 1-6, 2-1, 3-1, 3-2, 3-3, 3-4, 6-5, 7-4

3. **Aromatic substitution (electrophilic, nucleophilic, radical),** 6-1, 6-8, 7-1, 7-2, 7-6, 8-6, 9-1, 9-4, 9-7, 9-10

4. **Nucleophilic aliphatic substitution,** 1-9, 2-4, 6-3, 6-4, 7-4, 7-7, 8-1, 8-3, 8-5, 8-7, 9-12, 10-1, 10-5

5. **Elimination, extrusion, and decarboxylation,** 6-5, 7-4, 7-7, 7-9, 9-2, 11-2, 11-3, 11-4

6. **Electrophilic addition,** 6-3, 7-3, 9-3, 9-4, 9-5, 10-2, 10-4, 10-7, 10-8, 10-10, 10-12

7. **Nucleophilic addition,** 4-6, 6-2, 6-3, 8-1, 8-2, 9-6, 9-8, 9-9, 10-3, 10-5, 10-6, 10-9, 10-10, 10-11

8. **Pericyclic reactions,** 6-3, 7-8, 8-1, 8-3, 9-11, 11-1, 11-2, 11-3, 11-4, 11-5, 11-6, 11-7, 11-8, 11-9, 11-10

9. **Ionic rearrangements,** 6-5, 6-6, 7-4, 7-5, 9-12

10. **Reactive intermediates**
    A. **Carbocations,** 2-1, 4-5, 5-3, 6-4, 7-4, 7-7, 8-3, 8-4, 8-5, 9-3, 9-5, 9-7, 9-12
    B. **Carbanions and nitrogen anions,** 1-9, 5-2, 6-6, 7-5, 10-2
    C. **Free radicals, carbenes, and nitrenes,** 2-5, 2-6, 3-4, 6-3, 6-4, 6-7, 7-2

1
CONFIGURATIONAL AND CONFORMATIONAL ANALYSIS

PROBLEMS

1. Calculate the difference in enthalpy between the cis- and trans-decalins by counting the number of gauche-butane interactions. Use $\Delta \underline{H}^\circ = 0.8$ kcal mol^{-1} for each gauche interaction.

2. a. Draw the stable conformations of n-butane, 2-methylbutane, and 2,3-dimethylbutane in Newman projection. By how many gauche-butane interactions do the conformers differ for each molecule?

   b. The enthalpy differences between conformers are

n-butane	966 cal mol^{-1}
2-methylbutane	809
2,3-dimethylbutane	54

   Reconcile these observations with your gauche-butane count in (a). Use structural arguments.

3. a. Write out all possible conformational forms of 1-methyl-quinolizidine. Ignore enantiomers.

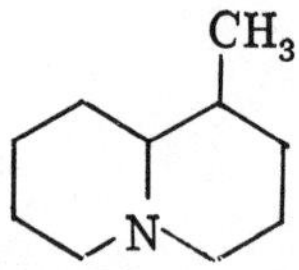

b. Which forms interconvert with which (without bond breaking)? Note that both ring reversal and nitrogen inversion can take place (in contrast to 1-methyldecalin).

c. Count the number of gauche-butane interactions for each structure in (a). In each set of interconverting forms, which one(s) is (are) formed?

4. Draw out all the diastereoisomers of perhydrophenanthrene and assess their approximate relative stability.

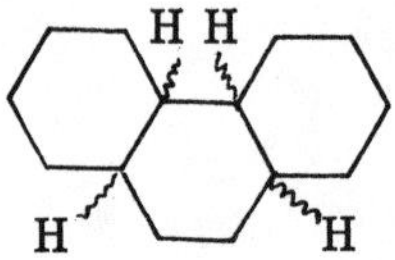

5. a. Suggest four reasonable conformations for cycloheptane (C_7H_{14}). Draw three-dimensional representations of them and discuss the energetic problems they may have (eclipsing strain, nonbonded interactions, etc.).

b. How many different substituent positions are there in the most stable form?

6. 2,3-Dibromobutane exists in meso and dl forms. If the diastereoisomers could be equilibrated chemically, what would be the equilibrium constant at 25 °C? Follow these steps:

a. Write down the three Newman projections for each form.

b. With the knowledge that the gauche CH_3/CH_3 interaction is 0.8 kcal mol^{-1}, Br/Br is 0.7, and CH_3/Br is 0.2, calculate the enthalpies for each conformer.

c. First for the meso, then for the dl, calculate the mole fractions N_i of each conformer from the formulas

$$\frac{N_i}{N_k} = e^{-(H_i - H_k)/RT}$$

$$\Sigma N_i = 1.0.$$

d. Calculate the overall enthalpies,

$$H_{meso} = \underset{i}{\Sigma} N_i H_i$$

$$H_{dl} = \underset{i}{\Sigma} N_i H_i .$$

e. Calculate the entropies of mixing, S_{meso} and S_{dl}, from the formula

$$S_{mix} = -R\underset{i}{\Sigma} N_i \ln N_i .$$

For the dl form, add $R \ln 2$, since each conformer has a nonsuperimposable mirror image. Symmetry numbers cancel.

f. Calculate G_{meso}, G_{dl}, ΔG, and K.

7. Characterize the indicated protons as being stereohomotopic, enantiotopic, or diastereotopic.

a.

b.

c.

d.

e.

f.

g.

h.

i.

j.

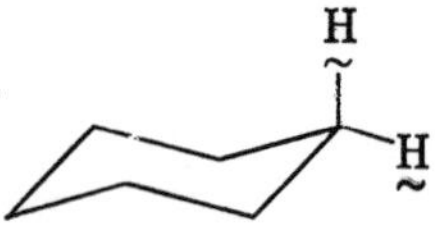

k.

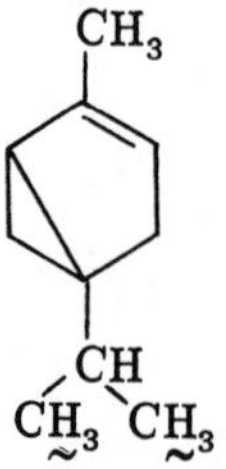

l.

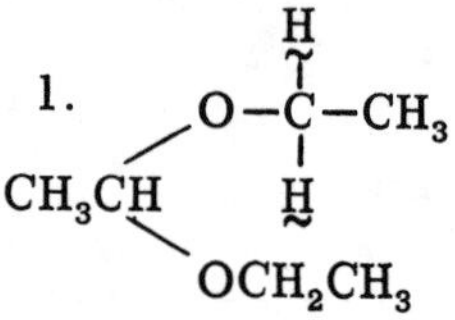

m.

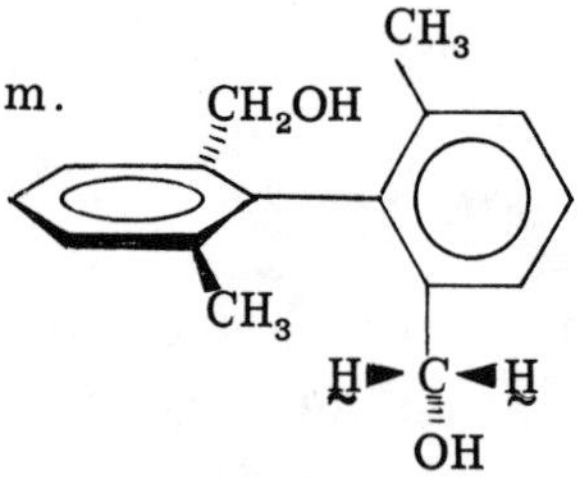

8. By steric arguments, account for the rate differences in the oxidation of the following alcohols to the corresponding ketones (CrO_3 in 89.7% HOAc that is 0.95 M with respect to H_2SO_4 at 25 °C).

$\underline{k}$, ℓ mol^{-1}sec^{-1}

A

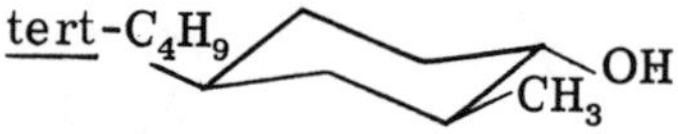

42

B

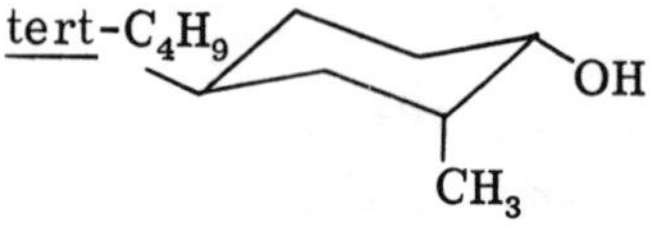

115

C

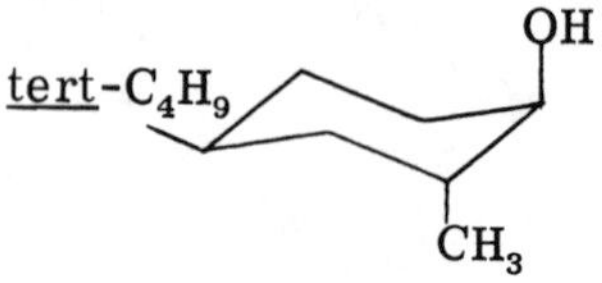

123

D

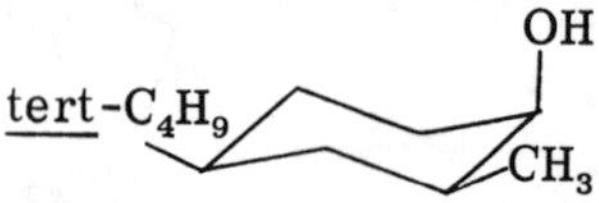

146

4

E 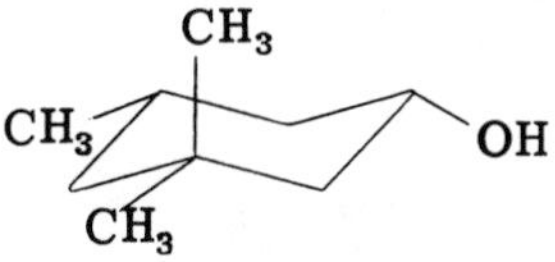   (25)[a]

F 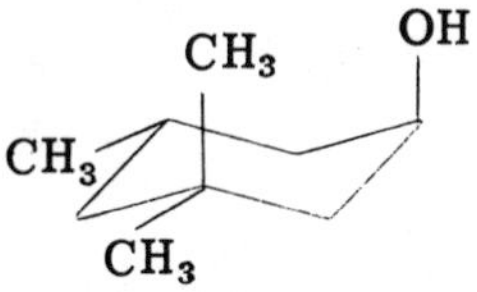   (828)[a]

[a]Approximated from data in 75% HOAc.

9. a. The syn α protons in N-nitrosopiperidine exchange with
deuterium 1000 times faster than the anti protons. (Syn and
anti refer here to the geometry with respect to the -N=O
group.) Of the two syn protons, the axial exchanges 100
times faster than the equatorial. Provide a stereoelectron-
ic explanation for these observations. The N-N=O functionality is planar. Hint: Look at the carbanion.

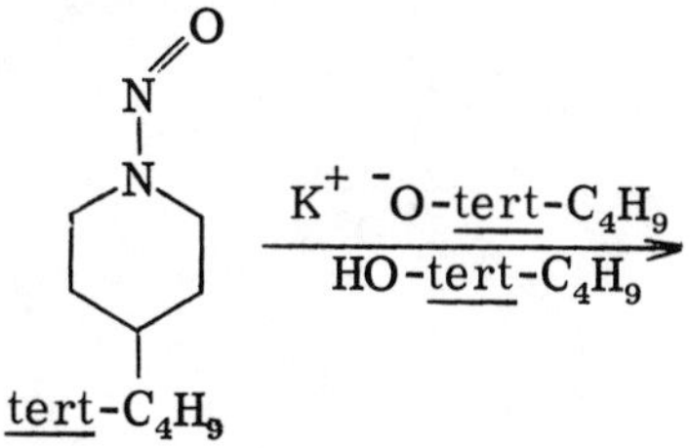

b. On the basis of the above observations, predict the exact
stereochemistry of alkylation of oxime O-methyl ethers.
Draw a three-dimensional structure of the product in the
reaction below and explain.

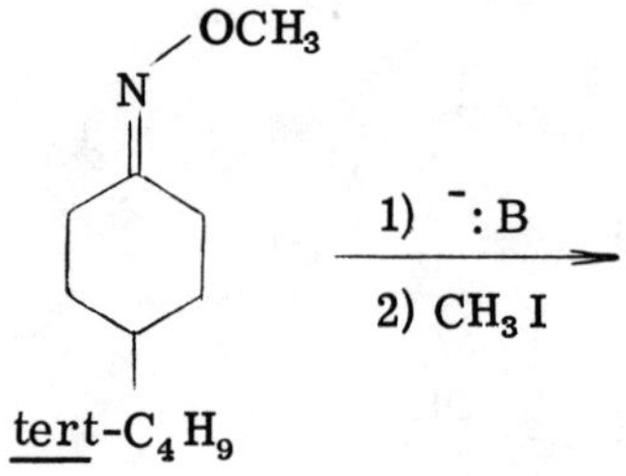

$$
\text{1) } {}^- : B
$$
$$
\text{2) } CH_3 I
$$

SOLUTIONS

1. It would be useful to follow these descriptions with molecular models. First consider the 1,2-dimethylcyclohexanes.

Whereas the trans form has only the one gauche interaction between the substituent methyl groups, the cis form has three, one between the methyls and one between the axial methyl and each of the gauche bonds within the ring (darkened bonds). Thus the difference is two gauche interactions, or about 1.6 kcal mol^{-1}.

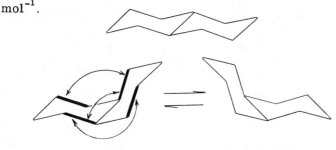

In trans-decalin, there are no gauche interactions from one ring to the other. One does not count gauche interactions within a given ring, because there are an equal number in all other forms and no energy differential results. In cis-decalin there are three gauche interactions, one between the axial bond in the right-hand ring and each of the gauche bonds in the left-hand ring and one between the axial bond in the left-hand ring (rotate the page to be convinced that the bond is indeed axial) and the gauche bonds in the right-hand ring. One interaction was counted twice, so that there are a total of three gauche-butane interactions in cis-decalin, as indicated. Thus the cis-trans enthalpy difference would be about 2.4 kcal mol^{-1}. It is useful to note that the trans form is entirely rigid, but the cis form undergoes double ring reversal.

Reference: E. L. Eliel, "Stereochemistry of Carbon Compounds," McGraw-Hill, New York, NY, 1962, pp 279-80.

2. a. n̲-butane

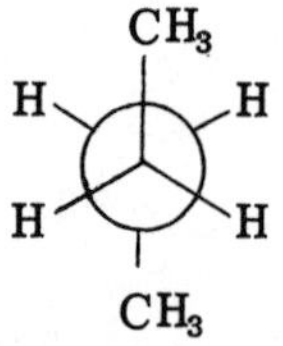

CH₃   CH₃

H   H   H   CH₃

H   H   H   H

CH₃   H

0 interactions 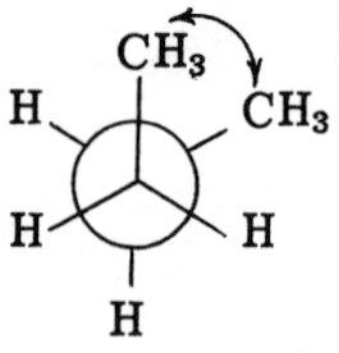   1 interaction

difference: 1 interaction

2-methylbutane

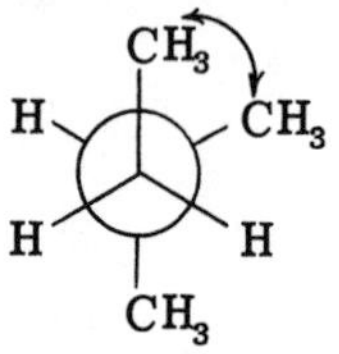

CH₃   CH₃

H   CH₃   CH₃   CH₃

H   H   H   H

CH₃   H

1 interaction 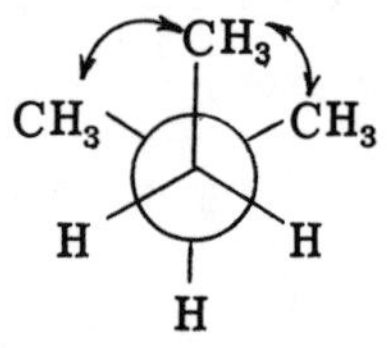   2 interactions

difference: 1 interaction

2,3-dimethylbutane

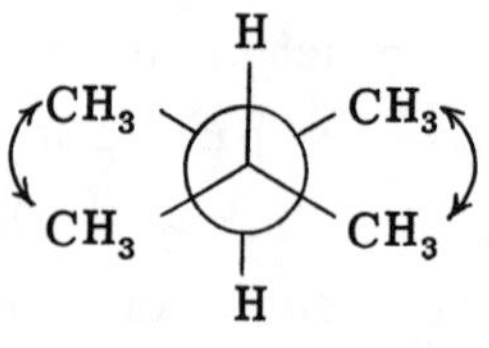

H   H

CH₃   CH₃   H   CH₃

CH₃   CH₃   CH₃   CH₃

H   CH₃

2 interactions 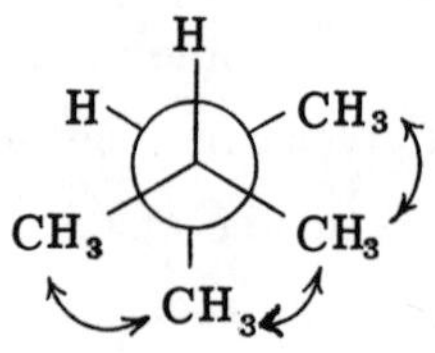   3 interactions

difference: 1 interaction

Thus for each molecule, the conformers differ by one gauche-butane interaction.

b. A n̲-butane

   B 2-methylbutane

C 2,3-dimethylbutane

Clearly, the simple concept of counting gauche-butane interactions does not carry over to highly crowded systems,
since $\Delta \underline{H}$ for one gauche-butane interaction is different in all
three systems and is particularly low in C. Some structural
alteration must serve to relieve the strain in C. The most
likely possibilities are torsional angle (C-C-C-C) and
valence angle (C-C-C) distortions. Even in the gauche form
of $\underline{n}$-butane (on the right above), the methyl-methyl torsional
angle is 66°, not 60°, and the CH_3-CH_2-CH_2 valence angle is
113.3°, compared to 111.9° in the trans form. As a result,
the gauche methyl carbons are 3.158 Å apart. In the unsymmetrical form (on the left) of 2-methylbutane (B), similar
adjustments can take place. The trans methyl-methyl angle
is 186.6°, not 180°, and the C1-C2-C3 angle is 115° (the
C2-C3-C4 angle cannot open without pushing C4 into the
methyl on C3). The resulting gauche methyl distance is
3.122 Å. No torsional adjustment is possible in the symmetrical form (on the right) of B, since rotation in either direction would push the gauche methyls into each other. The
C1-C2-C3 angle, however, is free to bend out, so that the
gauche methyl distance is 3.143 Å. The angle is opened up
an exceptional amount, since <u>two</u> gauche-butane interactions
are relieved. In the $\underline{C}_{2h}$ form (on the left) of 2,3-dimethyl-
butane (C) neither torsional nor valence angle distortion can
occur, since either distortion would push vicinal or geminal
methyls together (C1-C2-C3-C4 is 180°, C1-C2-C3 is
112.2°). As a result, the gauche methyl distance is excep-
tionally short, 3.037 Å. In the $\underline{C}_2$ form (on the right), torsional adjustment improves two and aggravates one gauche
interaction, with a net relief by opening the C1-C2-C3-C4
angle to 70.6°. Although there is some tightening of the
geminal methyls, valence angle opening (to 113.6°) relieves
<u>two</u> gauche-butane interactions. The resulting gauche methyl

distances are 3.119 and 3.148 Å. Thus the absence of strain relief in the $\underline{C}_{2h}$ form and the presence of torsional and valence angle adjustments in the $\underline{C}_2$ form offset the fact that the latter has one more gauche-butane interaction. The two forms have almost the same energy. This example illustrates the fact that counting gauche-butane interactions without considering other distortions can be misleading. Although this explanation utilized quantitative data from the literature, the same conclusions could have been reached by qualitative reasoning.

Reference: R.H. Boyd, $\underline{J}$. $\underline{Am}$. $\underline{Chem}$. $\underline{Soc}$., 97, 5353 (1975).

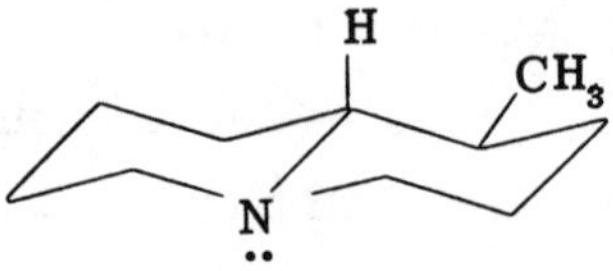

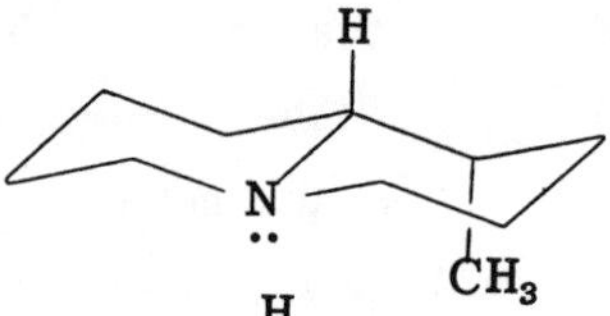

3. a.

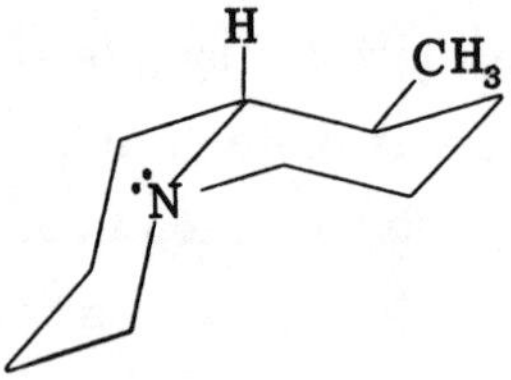

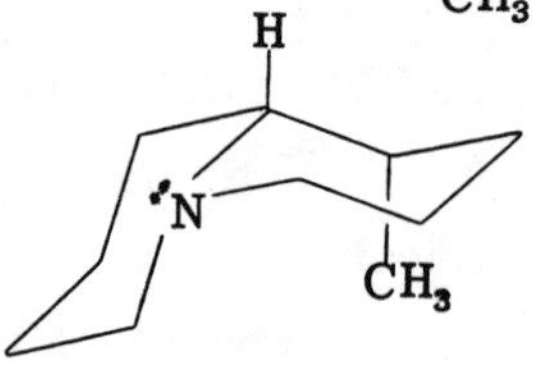

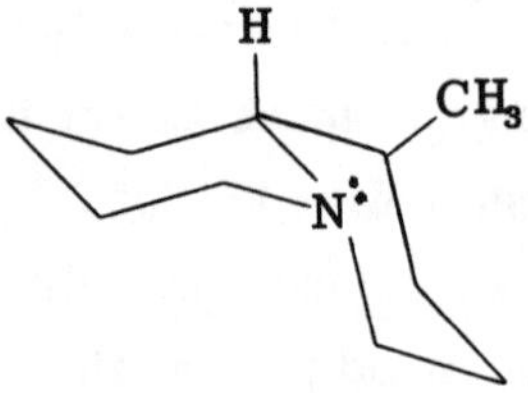

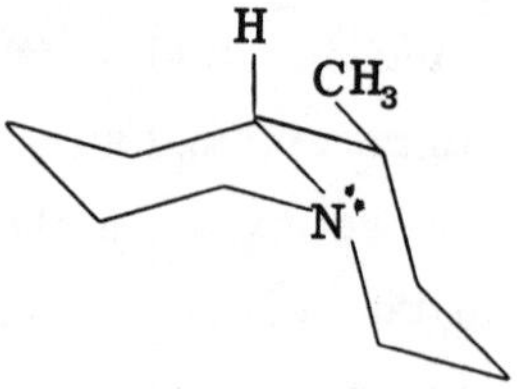

Two structural relationships can be altered, the stereochemistry of the ring fusion (cis or trans) and the axial or equatorial location of the 1-methyl group. For the trans ring junction, the methyl group can be either equatorial (A) or axial (D). The bridgehead proton is axial to both rings in A and D. There are four cis-fused structures. When the methyl group is equatorial, there is one form (B) in which the bridgehead proton is axial to the methyl-substituted ring (equatorial to the unsubstituted ring) and another (F) in which the bridgehead proton is equatorial to the methyl-substituted ring (axial to the unsubstituted ring). When the methyl group is axial, there is one form (C) in which the bridgehead proton is equatorial to the methyl-substituted ring (axial to the unsubstituted ring) and another (E) in which the bridgehead proton is axial to the methyl-substituted ring (equatorial to the unsubstituted ring). These relationships should be verified from models if they are not apparent from the drawings.

b. Double ring reversal (DRR) interconverts one cis form with another. Nitrogen inversion (NI) and single ring reversal (SRR) interconverts a cis form with a trans form. Thus B and C interconvert by ring reversal, as do E and F. Nitrogen inversion converts A to B or C and D to E or F. Two configurational sets (A, B, C and D, E, F) exist. Chirality has not been considered here.

$$\begin{array}{ccc} & \text{DRR} & \\ B & \rightleftharpoons & C \\ \text{NI, SRR} \searrow & A & \swarrow \text{NI, SRR} \end{array} \qquad \begin{array}{ccc} & \text{DRR} & \\ E & \rightleftharpoons & F \\ \text{NI, SRR} \searrow & D & \swarrow \text{NI, SRR} \end{array}$$

c.

Form	No. of gauche-butane interactions
A	1
B	4
C	6
D	3
E	6
F	4

To count the number of interactions, four considerations are important.

(i) Interactions of an axial 1-methyl group with the attached ring (two each in C, D, and E).

(ii) Interactions of the methyl group with the gauche bond in the other ring (one each in every case).

11

(iii) Interactions of an axial bond to nitrogen with a gauche bond in the other ring (two each in B, C, E, and F).

(iv) Interactions of an axial bond at the carbon bridgehead with a gauche bond in the other ring (two each in B, C, E, and F, but one has already been counted in (iii) for every case).

In the A⇌B⇌C set, A is clearly favored. In the D⇌E⇌F set, D is favored but F may contribute significantly.

References: E. L. Eliel, N. L. Allinger, S. J. Angyal, and G. A. Morrison, "Conformational Analysis," Wiley-Interscience, New York, NY, 1965, pp 252-53; T. M. Moynehan, K. Schofield, R. A. Y. Jones, and A. R. Katritzky, J. Chem. Soc., 2637 (1962).

4. In multiring systems, the diastereomers can first be displayed in planar projection by depicting all possible combinations of ring junction stereochemistry (cis or trans at each of the junctions) and of the relative stereochemistry of the end rings (syn or anti). There are six possibilities here, both junctions cis and the end rings syn or anti, both junctions trans and the end rings syn or anti, one junction each cis and trans and the end rings syn or anti. This procedure omits conformational diastereoisomers.

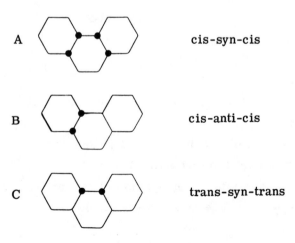

A   cis-syn-cis

B   cis-anti-cis

C   trans-syn-trans

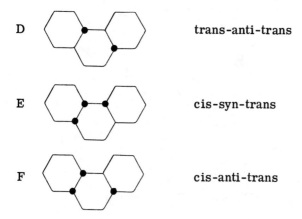

D   trans-anti-trans

E   cis-syn-trans

F   cis-anti-trans

It is noted that A and C are meso (achiral) and B, D, E, and F are chiral. The rings must now be transcribed into three dimensions.

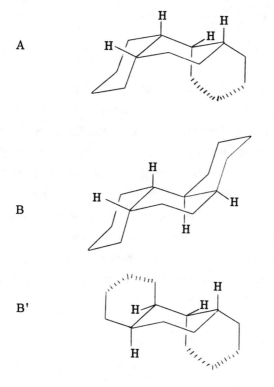

A

B

B'

13

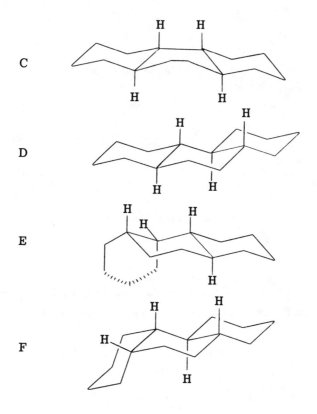

Normally, a chair six-membered ring is viewed in perspective side-on. Systems B, D, and E can be drawn entirely with that familiar perspective. One ring each in A and E and two rings in B' must be viewed face on, so that a somewhat awkward perspective results. Confusion can be decreased by examining molecular models. All the bridgehead protons have been drawn in. It is important to confirm that these protons retain the orientation defined by the planar depictions. Form C must have the central ring in a boat, because a chair form would have to possess the impossible property that one of the trans ring junctions have the substituents axial-axial.

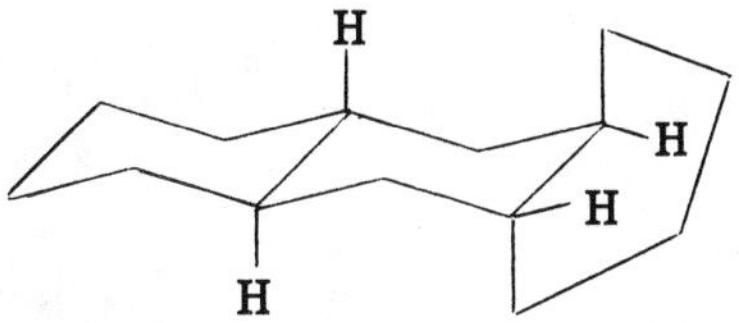

(impossible)

In searching for conformational multiplicities, one can eliminate all systems with even one trans ring junction (C, D, E, F). These species can only exist in one form with the trans junction equatorial-equatorial, since the axial-axial form is impossible (see above structure). Species A and B, however, have only cis ring junctions. Ring reversal of A gives a conformational isomer of identical structure (mirror image), so that no new form is produced. The meso condition is only fulfilled by rapid ring reversal. Ring reversal of B, however, gives a new form B' with a distinct structure. Thus there are seven, not six diastereoisomers of perhydrophenanthrene, although two can interconvert via ring reversal. Three principles can be utilized to approximate the relative order of stabilities.

(a)   Number of gauche-butane interactions: D (1), E (4), F(4), B' (6), B (7), A (7).

(b)   Disposition of two axial carbons syn to each other: B (0), B' (0), A (1).

(c)   Presence of a boat form: C

If the boat form is taken as the worst condition, the following relative order results (most to least stable):

D > E ~ F > B' > B > A > C

This approach supplies only a first-order approximation. Reference: M. Hanack, "Conformation Theory," Academic Press, New York, NY, 1965, pp 209-214.

5.   a.   The two most frequent answers offered by students are the chair and the boat, by analogy to cyclohexane.

Chair   Boat

Both these forms have eclipsing strain in the C4-C5 region.
The boat form also has some C2-C3 (C6-C7) eclipsing strain.
In addition, there are severe nonbonded interactions between
certain C-H pairs, in the chair between H3 and H6, H2 and
H7, and to a lesser extent H1 and H3/H6, in the boat between
H3 and H6, H2 and H7, and to a lesser extent H1 and H4/H5.
The net result of these interactions is that the chair form is
about 1 kcal mol^{-1} stabler than the boat form. Torsion about
C4-C5 in both the chair and the boat not only relieves the
eclipsing strain but also decreases the nonbonded interactions
somewhat. The resulting conformations are termed the
twist-chair and the twist-boat. Molecular models must be

Twist-chair   Twist-boat

examined to gain an appreciation of these shapes. With the
reduction of eclipsing and nonbonded interactions, the twist-
chair is about 2 kcal mol^{-1} stabler than the chair. The twist-
boat is only about 0.5 kcal mol^{-1} stabler than the boat.
Rapid interconversion of the boat to the twist-boat and the
chair to the twist-chair is termed pseudorotation. The chair
and boat forms are at energy maxima and the twist chair and
twist-boat forms at energy minima. Interconversion of the

16

chair family with the boat family requires 6-10 kcal mol^{-1}.

b.

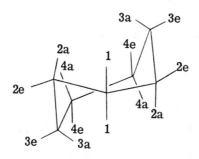

The stablest form, the twist-chair, possesses a $\underline{C}_2$ axis but lacks a plane of symmetry. Carbons 2, 3, and 4 each have distinct quasi-axial and quasi-equatorial substituent positions. Because of the $\underline{C}_2$ axis the two positions at C1 are equivalent. Thus there are seven distinct substituent positions, 1, 2a, 2e, 3a, 3e, 4a, 4e. Pseudorotation (twist-chair ⇌ chair) rapidly interconverts these positions. For a single methyl substituent, the order of positional preference is 2e ~ 3e ~ 4e > 1 > 4a > 2a ~ 3a. Geminal substituents would prefer the two equivalent 1 positions.

References: E. L. Eliel, N. L. Allinger, S. J. Angyal, and G. A. Morrison, "Conformational Analysis," Wiley-Interscience, New York, NY, 1965, pp 207-209; J. B. Hendrickson, J. Am. Chem. Soc., 83, 4537 (1961).

6. a.

meso:

a	b	c
CH$_3$	CH$_3$	CH$_3$
Br—⟨⟩—H	H—⟨⟩—CH$_3$	CH$_3$—⟨⟩—Br
H—⟨⟩—Br	H—⟨⟩—Br	H—⟨⟩—Br
CH$_3$	Br	H

dl:

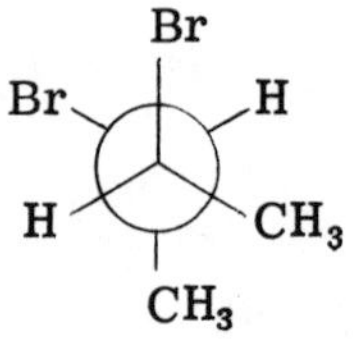

| d | e | f |

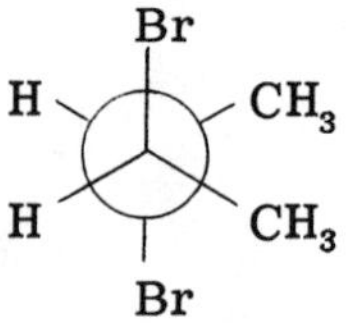

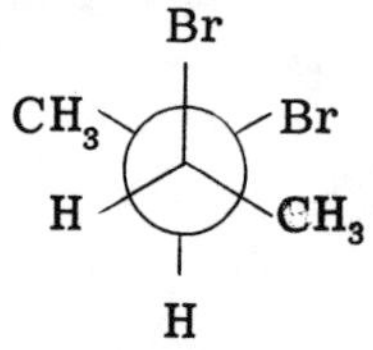

b. meso:

$$\underline{H}_a = 2 (0.2) = 0.4 \text{ kcal mol}^{-1}$$

$$\underline{H}_b = 0.8 + 0.2 + 0.7 = 1.7$$

$$\underline{H}_c = 0.8 + 0.2 + 0.7 = 1.7$$

dl:

$$\underline{H}_d = 0.8 + 0.7 = 1.5 \text{ kcal mol}^{-1}$$

$$\underline{H}_e = 0.8 + 2 (0.2) = 1.2$$

$$\underline{H}_f = 0.7 + 2 (0.2) = 1.1$$

These interactions are solvent dependent. Appreciably different results therefore might be expected in different solvents. In this calculation, other sources of strain, e.g., angle bending, are neglected. The results therefore are approximate.

c. meso: $\underline{N}_a = 0.82$ $\underline{N}_b = 0.09$ $\underline{N}_c = 0.09$

dl: $\underline{N}_d = 0.22$ $\underline{N}_e = 0.36$ $\underline{N}_f = 0.42$

d. $\underline{H}_{meso} = 0.82 (0.4) + 0.09 (1.7) + 0.09 (1.7)$
   $= 0.63 \text{ kcal mol}^{-1}$

   $\underline{H}_{dl} = 0.22 (1.5) + 0.36 (1.2) + 0.42 (1.1)$
   $= 1.22 \text{ kcal mol}^{-1}$

Thus enthalpy favors the meso form.

e. $\underline{S}_{meso} = -\underline{R}(0.82 \ln 0.82 + 0.09 \ln 0.09 + 0.09 \ln 0.09)$
   $= 1.18 \text{ eu}$

   $\underline{S}_{dl} = -\underline{R}(0.22 \ln 0.22 + 0.36 \ln 0.36$
   $+ 0.42 \ln 0.42) + \underline{R} \ln 2$

$$= 2.12 + 1.38$$
$$= 3.50 \text{ eu}$$

Thus entropy favors the dl form.

f. $\underline{G}_{meso}$ $= 630 - 298 \ (1.18)$
$$= 280 \text{ cal mol}^{-1}$$

$\underline{G}_{dl}$ $= 1220 - 298 \ (3.50)$
$$= 180 \text{ cal mol}^{-1}$$

The net favoring of the dl form means that entropy is the dominant factor in this equilibrium.

$$\Delta \underline{G} = \underline{G}_{meso} - \underline{G}_{dl}$$
$$= 100 \text{ cal mol}^{-1}$$

$\underline{K}$ $= \underline{e}^{-\Delta \underline{G}/\underline{R}\,\underline{T}}$
$$= 0.84 \text{ (favoring the dl form)}$$

Reference: J. A. Hirsch, "Concepts in Theoretical Organic Chemistry," Allyn and Bacon, Boston, MA, 1974, pp 244-46.

7. Stereohomotopic atoms or groups may be interconverted by rotational symmetry. A simple test involves replacement of the atoms in question respectively by another isotope of the same element. Respective replacement of stereohomotopic atoms produces identical molecules. Thus the protons in difluoromethane are stereohomotopic.

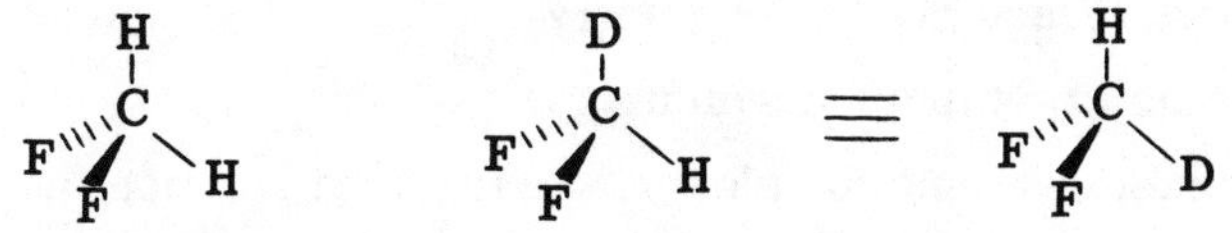

Enantiotopic atoms or groups may be interconverted by a symmetry operation other than rotation, e.g., mirror reflection. Respective replacement of enantiotopic atoms produces enantiomeric molecules, as in chlorofluoromethane. Diastereotopic

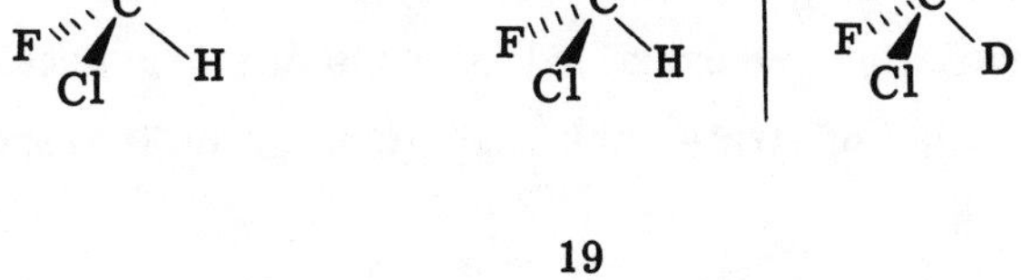

atoms or groups differ from one another only in their relationship in space but cannot be interconverted by any symmetry
operation. Respective replacement of diastereotopic atoms
produces diastereomeric molecules, as in 1-bromo-1-chloro-2-
fluoroethane.

a. Stereohomotopic ($\underline{C}_2$ axis).
b. Diastereotopic (no symmetry).
c. Enantiotopic (mirror plane in the plane of the benzene ring).
d. Diastereotopic (no symmetry).
e. Enantiotopic (mirror plane through the para substituents).
f. Stereohomotopic ($\underline{C}_2$ axis).
g. Diastereotopic (no symmetry).
h. Enantiotopic (mirror plane).
i. Enantiotopic (mirror plane).
j. Diastereotopic for slow ring reversal (no symmetry).
   Enantiotopic for fast ring reversal on the time scale (NMR
   or laboratory) of the experiment (average mirror plane).
k. Diastereotopic (no symmetry).
l. Diastereotopic (no symmetry).
m. Diastereotopic for slow ring-ring bond rotation (no symmetry). Enantiotopic for fast rotation (average mirror plane).

Reference: J. B. Lambert, H. F. Shurvell, L. Verbit, R. G.
Cooks, and G. H. Stout, "Organic Structural Analysis,"
Macmillan, New York, NY, 1976, pp 52-56.

8. These rates are influenced by both steric and polar factors, but
    the polar component can be neglected by comparing a pair of
    substrates with identical but sterically distinct substituents.
    The relative rates are explained in terms of steric acceleration
    caused by relief of strain as the sp^3 HCOH group is transformed

into an sp^2 C=O group. The transition state must be product-like for this strain relief to provide the dominant effect. Although this point has been questioned, it has recently been shown that the rates of CrO_3 oxidation are directly proportional to the strain energy of secondary alcohols (compared to the carbonyl product) for a wide variety of structures. For the substrates illustrated, the relative rates can be explained by three considerations. The 4-tert-butyl and the 5-methyl groups only serve as conformational anchors.

a. An axial hydroxyl group oxidizes more rapidly than equatorial hydroxyl (the factor is normally 3-5), as in D/A. The repulsive syn-axial H/OH interactions are removed by oxidation.

b. A substituent that is gauche to OH also causes an acceleration. Thus B (with relief of a gauche CH_3/OH interaction) oxidizes almost as rapidly as C, even though the B hydroxyl is equatorial and the C is axial. The gauche interaction is not equivalent in every case. The e/e interaction in A can be relieved by torsion, but the e/a interaction in B cannot since the CH_3 group would be pushed further toward the syn-axial 3,5 hydrogens. Thus relief of the gauche CH_3/OH interaction in A provides less acceleration than in B or D.

c. Relief of a syn-axial CH_3/OH interaction provides a very large acceleration, as in F/E.

References: J. C. Richer, L. A. Pilato, and E. L. Eliel, Chem. Ind. (London), 2007 (1961); F. Šipoš, J. Krupička, M. Tichý, and J. Sicher, Collect. Czech. Chem. Commun., 27, 2079 (1962); P. Müller and J.-C. Perlberger, J. Am. Chem. Soc., 98, 8407 (1976).

9. a. Syn exchange is preferred for the same reason that the cis
form of the butadiene dianion is stabler than the trans form.

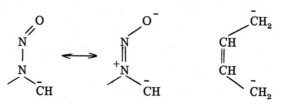

Steric and electrostatic factors would presumably favor the
transoid form (anti exchange), so that an alternative, electronic effect must be operative. One possible explanation
is that the termini of the dianion exert an attractive force
on each other in the cisoid form. Possible causes include
net $1,4 \pi$ bonding and lone pair-lone pair $n-\sigma^*$ interactions.
At least for 1,2-difluoroethylene (electronically similar to
the butadiene dianion), this explanation seems unlikely,
since the C=C-F angles in the favored cis form are larger,
not smaller, than in the trans form. Bingham has pointed
out that whereas electron delocalization favors zigzag
(trans) structures, cases in which delocalization is energe-
tically unfavorable favor nonzigzag (cis) structures. In the
delocalized butadiene dianion, electrons are present in two
bonding and one antibonding orbitals. The
filled antibonding orbital renders delocalization unfavorable, so the cis conformation, which minimizes delocalization, is
preferred. In the N-nitrosamine, transi-
tion state delocalization is minimized in syn exchange. The
syn (cisoid) preference is closely related to the gauche
effect, whereby two polar or lone pair-bearing groups prefer a gauche over a trans orientation. The proper interpretation of all these phenomena lies either in attractive
orbital interactions, inhibited delocalization, or an as yet
unconsidered phenomenon. The syn-axial protons exchange
more rapidly than the syn-equatorial protons because the

axial C-H bond lines up well with the π framework.

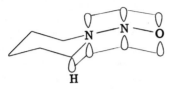

b. This question may be answered without reference to the theoretical explanations in (a). One expects empirically from the above result that the methyl group will replace the syn-axial proton.

$\underline{\text{tert-C}_4\text{H}_9}$ ⟍⟋ N ⟍OCH$_3$ H CH$_3$

References: R. R. Fraser and L. K. Ng, J. Am. Chem. Soc., 98, 5895 (1976); R.R. Fraser and K. L. Dhawan, J. Chem. Soc., Chem. Comm., 674 (1976); R. C. Bingham, J. Am. Chem. Soc., 98, 535 (1976).

ALSO SEE PROBLEMS 2-1, 3-1, 3-2, 3-3, 3-4, 4-4, 4-5, 4-8, 6-5, 7-4, 7-5.

2

ORGANOTHERMOCHEMISTRY

1. Related subject: conformational analysis

    The equilibrium between diamantan-1-ol (A) and -4-ol (B) is
    established in 98% H_2SO_4 (probably by an intermolecular hydride
    shift).

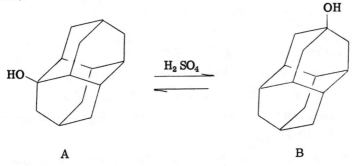

A   B

a. For both compounds, characterize the hydroxyl group as
    axial or equatorial for each of the rings in which it is a sub-
    stituent. Which isomer (A or B) is enthalpically favored?

b. Which isomer is entropically favored and why?

c. Experimentally, B is favored (56/44) at 273 °K, the isomers are equally favored (50/50) at 321 °K, and A is favored (35/65) at 473 °K. Explain these observations in terms of $\Delta \underline{H}°$ and $\Delta \underline{S}°$.

2. Alcohols such as ethanol and carboxylic acids such as acetic acid both form hydrogen bonds in the vapor phase. Only the acids, however, normally exist as hydrogen-bonded dimers in the vapor. Explain.

3. a. The empirical resonance energy (ERE, the energy difference between the π electron-delocalized species and the localized version with alternating bond lengths) of benzene is -36 kcal mol^{-1}. What then should the ERE of biphenylene be? The observed ERE is -10 kcal mol^{-1}. Account for the discrepancy between observed and expected.

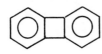

(No valence structures are meant to be implied by this representation.)

b. Azulene has a heat of hydrogenation of -100 kcal mol^{-1},

naphthalene of -82. Calculate the ERE's of azulene and naphthalene, given that the heat of hydrogenation of cyclopentene is -26, cyclohexene -28.6, and cycloheptene -28 kcal mol^{-1}. Comment on the differences between azulene and naphthalene.

4. Measurement of the heats of methylation of two compounds that give a common product is a useful procedure for determining relative stabilities. Consider the following data (kcal mol^{-1}) (referred to the vapor phase).

A $\xrightarrow[\text{(-32.6)}]{\text{CH}_3\text{FSO}_3}$ $\xleftarrow[\text{(-40.2)}]{\text{CH}_3\text{FSO}_3}$ B

C $\xrightarrow[\text{-48.3}]{\text{CH}_3\text{FSO}_3}$ $\xleftarrow[\text{-46.2}]{\text{CH}_3\text{FSO}_3}$ D

a. Which is the stabler of the isomer pair A, B and by how much? Which is the stabler of C, D?

b. The pair C, D can be used as a model for the localized-bond version of A, B. Calculate the difference in empirical resonance energy between A and B (see problem 2-3 for the definition of the ERE). Which (A or B) is more stabilized according to this criterion?

c. Summarize your answers to (a) and (b) by an energy level diagram.

5. The bond dissociation energy (BDE) is the energy for dissociation of a specific chemical bond to cleave a molecule into two fragments. The contributing bond energy (CBE) is the contribution from that bond to the atomization of the entire molecule. The difference between the BDE and the CBE for a bond has been called the reorganization energy, $\underline{E}_R$ = BDE - CBE.

a. For methane, the BDE is 103.2 kcal mol^{-1} and the CBE is 99.1. What is the significance of the sign and magnitude of $\underline{E}_R$?

b. For the methyl C-H bond of toluene, the BDE is 85 kcal mol^{-1} and the CBE is 98.7. What is the significance of the sign and magnitude of $\underline{E}_R$? Comment on the change in sign with respect to methane.

c. Just because there is only one of a given bond in a molecule does not imply that BDE = CBE, i.e., that $\underline{E}_{\underline{R}}$ = 0. Using the concept of reorganization energy, explain why, with the O-O bond of hydrogen peroxide as the example (BDE = 51.2 kcal mol^{-1}, CBE = 32.9).

6. The π bond strength for the carbonyl group is the difference between the energy of C=O with and without π overlap. One approach to evaluating this quantity calculates 90-93 kcal mol^{-1} from the difference between analogous C=O double bond and C-O single bond energies. A second approach gives 76-78 kcal mol^{-1} from the difference in bond dissociation energy (BDE; see question 2-5 for a definition) for the O-H bond in an alcohol HR$_2$CO-H and in its analogous radical R$_2\overset{\centerdot}{C}$O-H.

a. What is the rationale for the second approach?

b. The difference in BDE for the C-H bond in ROCR$_2$-H and in CH$_3$-H is 10-14 kcal mol^{-1} (methane higher). Why?

c. From the results in (b), indicate which of the two approaches to determine the carbonyl π bond strength is incorrect and point out the fallacy.

SOLUTIONS

1. a.

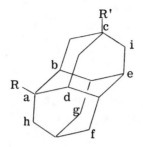

For isomer A (R = OH, R' = H), HO is axial in ring abcd and equatorial in rings abde, abgh, and adfh. For isomer B (R = H, R' = OH), HO is equatorial in all rings (abcd, bcei, cdei). Because of the one axial orientation in A, B is favored enthalpically ($\Delta \underline{H}^{\circ}$ is negative as written). The measured $\Delta \underline{H}^{\circ}$ (-1.1 kcal mol^{-1}) is not far from the $\underline{A}$ value for hydroxyl.

b. Isomer A has $\underline{C}_{\underline{s}}$ symmetry (a mirror plane but no axes, symmetry number of 1). Isomer B has $\underline{C}_{3\underline{v}}$ symmetry (a threefold axis through the C-OH bond, symmetry number of 3). Therefore isomer A is favored entropically, since it is less symmetrical ($\Delta \underline{S}^{\circ}$ is negative as written). The measured $\Delta \underline{S}^{\circ}$ (-3.4 eu) probably derives almost entirely from these symmetry differences, since both isomers are rigid. There might be a small but equal contribution for each isomer from C-OH rotation. The calculated symmetry factor ($-\underline{R} \ln 3 = -2.18$ eu) is not far from the observed entropy difference.

c. $\Delta \underline{G}^{\circ} = \Delta \underline{H}^{\circ} - T\Delta \underline{S}^{\circ} = -\underline{R}\,\underline{T} \ln \underline{K}$
   At 273 °K, B is favored because $|\Delta \underline{H}^{\circ}| > \underline{T} \,|\Delta \underline{S}^{\circ}|$ (enthalpy controlled). At 321 °K, the isomers are equally favored; $\Delta \underline{G}^{\circ} = 0$ because $\Delta \underline{H}^{\circ} = \underline{T} \Delta \underline{S}^{\circ}$. At 473 °K, A is favored because $\Delta \underline{H}^{\circ} < \underline{T} \Delta \underline{S}^{\circ}$ (entropy controlled).

Reference: D. E. Johnston, M.A. McKervey, and J. J. Rooney, J. Chem. Soc., Chem. Commun., 29 (1972).

2. An acid can form two hydrogen bonds per dimer, whereas an alcohol can form only one.

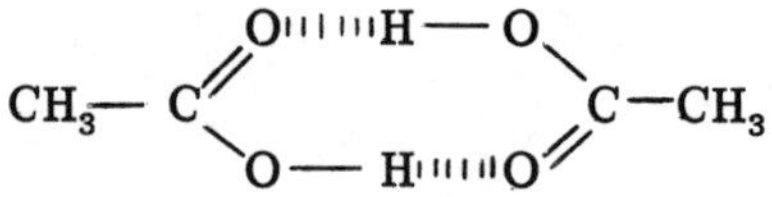

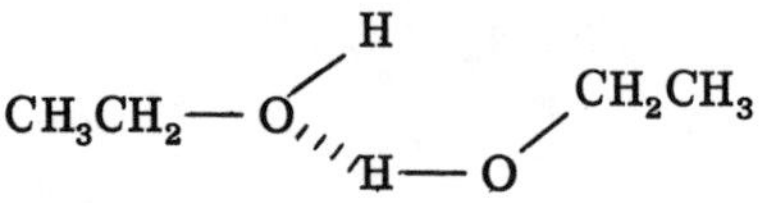

$$CH_3- C \overset{O\cdots\cdots H-O}{\underset{O-H\cdots\cdots O}{\diagup}} C-CH_3$$

$$CH_3CH_2- O \overset{H}{\underset{H-O}{\diagup}} CH_2CH_3$$

Any monomer $\rightleftharpoons$ dimer equilibrium (2A $\rightleftharpoons$ B) involves loss of translational degrees of freedom. Consequently, a dimerization reaction is disfavored entropically and will not occur unless the energy of the newly formed bonds in the dimer is sufficient to outweigh the loss of translational degrees of freedom. The single hydrogen bond in the alcohol dimer must be insufficient (entropy outweighs enthalpy), but the two bonds in the acid are sufficient (enthalpy outweighs entropy). Typically, a hydrogen bond might be worth 5 kcal mol^{-1}, and the entropy loss in the monomer $\longrightarrow$ dimer conversion might be 30 eu. For the alcohol

$$\Delta \underline{G}° = -5000 - 300 \times (-30) = 4000 \text{ cal mol}^{-1}$$

(monomer favored).

For the acid

$$\Delta \underline{G}° = -10,000 - 300 \times (-30) = -1000 \text{ cal mol}^{-1}$$

(dimer favored).

One must also consider that the hydrogen bonds do not have the same energy in the two systems, but this effect is probably subsidiary.

Reference: K. B. Wiberg, "Physical Organic Chemistry," Wiley, New York, NY, 1964, p 221.

3. Various terms concerned with the so-called resonance energy have been used, so it is important to appreciate the difference between an empirical resonance energy (ERE) and a vertical resonance energy (VRE). Benzene (B) has equal C-C bond

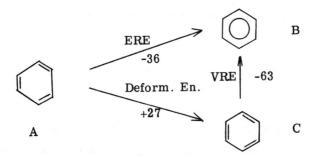

lengths and full π electron delocalization. A hypothetical cyclo-
hexa-1,3,5-triene (A) should have alternating bond lengths and
localized double bonds. These bonds resemble that of cyclohexene, so that the heat of hydrogenation of A should be three
times that of cyclohexene $(3 \times (-28.6) = -85.8$ kcal mol^{-1}). The
difference between this value and the observed value (-49.8) is
the ERE (-36 kcal mol^{-1}). The figure is empirical because it
includes not only the energy improvement that accompanies electron delocalization but also the energy cost to deform the bonds
to equality. This latter deformation energy has been calculated
to be about 27 kcal mol^{-1}, so the actual gain in stability is about
-63 kcal mol^{-1} (-36 - 27). This final figure is the VRE, or the
energy improvement in benzene (B) due entirely to delocalization
of electrons from a hypothetical localized, equal-bond form (C).
Because the ERE is frequently measurable, it is the normally
quoted quantity. Some limitations to this approach are discussed
in the references.

a. The two benzene rings each should have an ERE of -36 kcal
   mol^{-1}, for a total of -72. The value should be even greater
   than -72 because of conjugation between the two rings. The
   double bonds are nearly localized, i.e., like A above. Without delocalization, the rings possess no extra stability. Two
   reasons can be given for localized bonds.

(i) The molecule assumes a localized structure D in order to avoid the cyclobutadiene structure E. Because of the

D   E

antiaromaticity of the central ring in E ($4\underline{n}$ π electrons), this resonance structure will not contribute significantly to the composite. The net result is a localized form approximating D.

(ii) The highly strained four-membered ring causes the molecule to be nonplanar. The buckling of the benzene rings inhibits delocalization.

Of these two factors, (i) is probably more important.

b. For naphthalene $5(-28.6) = -143$ kcal mol^{-1}

$$\Delta\underline{H}(H_2) \quad = -82$$

$$\overline{ERE \quad = -61}$$

For azulene

$$3(-28) + 2(-26) = -136$$

$$= -100$$

$$\overline{ERE \quad = -36}$$

Since the molecules are isomers, the nonalternate azulene clearly has less aromatic stabilization. It should be emphasized that these calculations are fraught with limitations. The choice of the localized models, e.g., cyclopentene, is not rigorously correct. Whereas all bonds in azulene and naphthalene are sp^2-sp^2, those attached to the double bond in the hydrogenation models are sp^3-sp^2. Although these limitations detract from the quantitative rigor in calculations of this sort, they do not appreciably vitiate qualitative interpretations based on them.

References: A. Streitwieser, Jr., "Molecular Orbital Theory for Organic Chemists," Wiley, New York, NY, 1961, pp 237-47; R. D. Gilliom, "Introduction to Physical Organic Chemistry," Addison-Wesley, Reading, MA, 1970, pp 77-79.

4. a. The difference for A, B is

$$-40.2 - (-32.6) = -7.6 \text{ kcal mol}^{-1} \text{ (A stabler)}.$$

The difference for C, D is

$$-46.2 - (-48.3) = 2.1 \text{ (D stabler)}.$$

b. The ground state difference is $7.6 \text{ kcal mol}^{-1}$, and to this must be added the difference between the localized models (2.1). See part (c) below. Thus the difference in ERE's is $2.1 + 7.6 = 9.7 \text{ kcal mol}^{-1}$. 2-Methylthiopyridine (A) has a larger ERE than 1-methyl-2-thiopyridone (B) by this amount.

c.

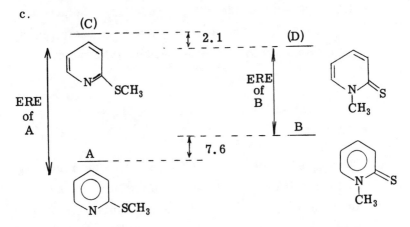

Compounds C and D are used as models for the localized versions of A and B, as illustrated. The actual ERE's cannot be evaluated, since C and D contain only one of the three localized double bonds. Although the models are imperfect, the difference in ERE may be fairly good.

References: P. Beak and J. T. Lee, Jr., J. Org. Chem., 34, 2125 (1969); P. Beak, D. S. Mueller, and J. Lee, J. Am. Chem. Soc., 96, 3867 (1974). Liberties have been taken with the original data for A and B. Although the heats of methylation were measured for C and D, only the enthalpy difference was reported for A and B. Hence the figures under the arrows (-32.6, -40.2) are fabricated for the sake of the problem (their difference is as reported), so they are enclosed in parentheses.

5. a. Because BDE > CBE for CH_4 (E_R = 4.1), more energy is required to break the first bond (CH_3-H) than the average of the four C-H bonds [CBE = $\frac{1}{4} \Delta H_f(CH_4)$]. The magnitude of E_R is small because the four bonds to be broken serially (CH_3-H, $\cdot CH_2$-H, :CH-H, :$\dot{C}$-H) are of similar strength. The principal contribution to the smaller value for the CBE is the fact that the last step, to form elemental carbon, :$\dot{C}$-H $\longrightarrow$:C + H$\cdot$, is of relatively low energy.

   b. In contrast to methane, less energy is required to break the $C_6H_5CH_2$-H bond than the average of the CH_3 bonds (E_R = -13.7). The magnitude of E_R is quite large, because cleavage of the first C-H bond is a relatively low energy process that forms the resonance-stabilized benzylic radical.

   c. Cleavage of the HO-OH bond does not leave the remaining O-H bonds unaffected. Cleavage of any bond in a molecule alters the energetics of the remaining bonds. Since the HO-OH BDE is larger by 18.3 kcal mol^{-1} than the CBE, the O-O bond must exert a weakening influence on the H-O bonds within the molecule, by $\frac{1}{2} E_R$ = 9.2. From the opposite point of view, the H-O bonds strengthen the O-O bond. Cleavage of the O-O bonds therefore strengthens the O-H bonds [BDE(HOO-H) = 90 kcal mol^{-1}, BDE ($\cdot$O-H) = 102]. The interdependence of bond strengths on neighboring bonds means that the HO-OH BDE cannot be taken as the standard O-O single bond energy. One would expect a different value, for example, for CH_3O-OCH_3.

Reference: R. T. Sanderson, J. Am. Chem. Soc., 97, 1367 (1975).

6. a. The BDE's correspond to the following processes.

$$HR_2C-O-H \xrightarrow{-H\cdot} HR_2C-\dot{O} \qquad\qquad A$$

$$R_2\dot{C}-O-H \xrightarrow{-H\cdot} R_2\dot{C}-\dot{O} \longleftrightarrow R_2C{=}O \qquad B$$

In process A, the resultant radical has no π overlap. In process B, the radical on oxygen and the already existing radical on carbon become paired through π overlap. The BDE of B should be lower than that of A by exactly the CO π bond energy, since B recovers the π bond energy but A does not.

b. $$R'OCR_2-H \xrightarrow{-H\cdot} R'O-\dot{C}R_2 \longleftrightarrow R'\dot{C}{=}CR_2 \quad C$$

$$CH_3-H \xrightarrow{-H\cdot} \cdot CH_3 \qquad\qquad D$$

In methane (process D), there is no π overlap to stabilize the radical. The lower BDE of C indicates that there must be at least partial C–O double bonding ($\dot{C}$-O $\longleftrightarrow$ C=$\dot{O}$) in the radical on the right side of the equation. Although the 10-14 kcal mol^{-1} stabilization may be in part due to the polar nature of the RO group, the predominant effect is most likely resonance. The BDE in process C is lower because the radical is better stabilized than in D.

c. The radical on the left side of process B in part (a) is the same as the radical on the right side of C in part (b). The results in (b) indicate that this radical is stabilized to some extent by a resonance interaction. Thus process B does not represent full development of a π bond. To the extent that the radical on the left of B already has some π overlap, this approach underestimates the carbonyl π bond energy. Thus the second method appears to be in error, and the value of 90-93 kcal mol^{-1} should be considered more likely.

Reference: R. A. Firestone, Tetrahedron Lett., 735 (1976).

ALSO SEE PROBLEMS 3-4, 3-6, 4-7.

3
EQUILIBRIUM SOLVENT
AND ISOTOPE EFFECTS

PROBLEMS

1. Related subject: conformational analysis

Consider the following equilibrium data for 2-halocyclohexanones in two different solvents.

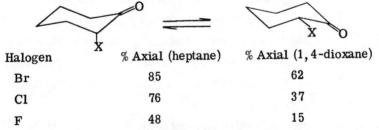

Halogen	% Axial (heptane)	% Axial (1, 4-dioxane)
Br	85	62
Cl	76	37
F	48	15

a. Explain the increase in the percent of axial conformer on going from 1, 4-dioxane to heptane as solvent.

b. There is a net axial preference for bromine in both solvents, but a net equatorial preference for fluorine. Explain the trend from bromine to fluorine.

2. Related subject: conformational analysis

Consider the ring reversal equilibrium for 5-substituted 1,3-dioxanes. The free energy differences (kcal mol^{-1}) have been

measured as a function of solvent (a negative $\Delta \underline{G}^{\circ}$ implies an equatorial preference). For monosubstituted cyclohexanes, the analogous data (relatively insensitive to solvent changes) are F (-0.15), Cl (-0.43), Br (-0.38), and OCH$_3$ (-0.55).

Solvent (ε)	F	Cl	Br	OCH$_3$
CCl$_4$ (2.24)	0.36	-1.40	-1.71	-0.90
(C$_2$H$_5$)$_2$O (4.33)	0.62	-1.26	-1.45	-0.83
CHCl$_3$ (4.81)	0.87	-0.94	-1.35	-0.16
CH$_3$CN (37.5)	1.22	-0.25	-0.68	0.01

a. Explain the direction of the solvent effect.

b. Explain the lower equatorial preference (more positive $\Delta \underline{G}^{\circ}$) for a given substituent in the 1,3-dioxanes dissolved in more polar solvents, compared to cyclohexanes.

c. Fluorine prefers the axial position in all solvents, whereas the remaining substituents exhibit a general preference for the equatorial position. The change in free energy differences ($\Delta \Delta \underline{G}^{\circ}$) is larger for F than for the other substituents. Explain the especially strong axial preference for 5-F in 1,3-dioxane.

3. Related subject: conformational analysis

The $\underline{A}$ value is a measure of steric interactions within the cyclohexane framework and is not a universal measure of substituent size. Thus alterations of the ring are expected to affect the $\underline{A}$-value concept. Consider the conformational equilibrium for 3-substituted $\underline{exo}$-methylenecyclohexanes.

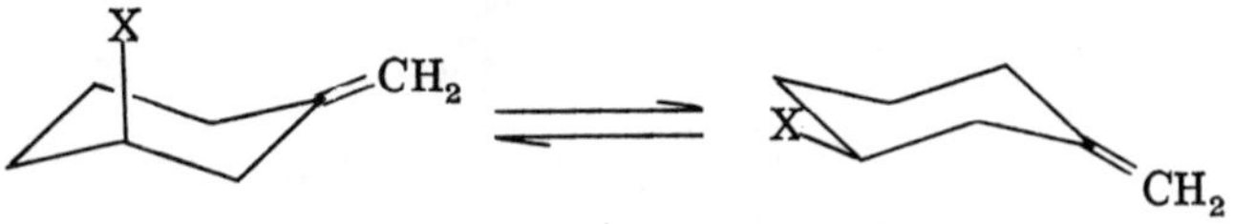

X	$-\Delta \underline{G}°$ (CF_2Cl_2)	$-\Delta \underline{G}°$ $(CHFCl_2)$	$\underline{A}$ value
OCH_3	0.80	0.11	0.55
OH	1.12	0.69	0.80
OAc	0.61	0.38	0.71
SCH_3	1.22	0.65	1.07
CH_3	0.6	0.5	1.8

a. Without regard to the data, describe your expectations for the relative value of $\Delta \underline{G}°$ in this equilibrium with respect to that in the parent cyclohexyl system (the $\underline{A}$ value).

b. The more nearly nonpolar solvent (CF_2Cl_2) should better reflect intramolecular interactions (as in gas phase equilibria). In CF_2Cl_2, $-\Delta \underline{G}°$ (except for OAc) is larger than $\underline{A}$, i.e., there is more equatorial conformer in the exo-methylene system, for OCH_3, OH, and SCH_3, but $-\Delta \underline{G}°$ is smaller for CH_3. Why?

c. In $CHFCl_2$, $-\Delta \underline{G}°$ is smaller than $\underline{A}$ for every case. Why?

d. Suggest a reason why $-\Delta \underline{G}°$ for OAc is smaller than $\underline{A}$ in both solvents (contrast OCH_3).

4. Related subjects: conformational analysis, organothermochemistry

A rotational mechanism for the cis-trans isomerization of azobenzene involves breaking the π bond, rotation, and reformation (the lone pairs remain in sp^2 orbitals):

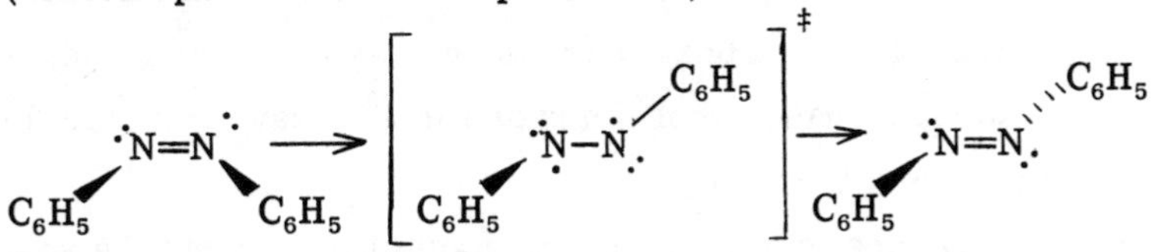

An inversional mechanism involves rehybridization of one N lone pair ($sp^2 \longrightarrow p$) and of the N orbital to C_6H_5 ($sp^2 \longrightarrow sp$).

Transfer of cis-azobenzene from cyclohexane to cyclohexanone (as solvent) is exothermic by 3.0 kcal mol^{-1} (measured by heats of solution). The enthalpy of activation for cis to trans isomerization is 22.0 kcal mol^{-1} in cyclohexane and 26.7 in cyclohexanone.

a. Draw and label an energy diagram that shows the ground state of cis-azobenzene and the transition state to isomerization for both solvents.

b. Arguing from the observed solvent effects and using the diagram, determine which is the mechanism for cis-trans isomerization of azobenzene.

5. In H_2O, acetic acid ($\underline{K}_a$ = 1.8 x 10^{-5}) is about three times more acidic than the pyridinium ion, $\langle O \rangle \overset{+}{N}-H$ ($\underline{K}_a$ = 6.2 x 10^{-6}. In CH_3OH (ε = 34), the relative acidities are reversed. The pyridinium ion dissociates about the same amount in both solvents ($\underline{K}_a$ = 2.8 x 10^{-6} in CH_3OH), but acetic acid dissociates almost 10^5 times less in CH_3OH ($\underline{K}_a$ = 2.2 x 10^{-10}). Thus in CH_3OH, the pyridinium ion is some 10^4 more acidic than acetic acid. Explain the reversal.

6. Related subjects: organothermochemistry, electronic effects on equilibria

a. In H_2O, phenol ($p\underline{K}_a$ ~ 10) is about six $p\underline{K}_a$ units less acidic than aliphatic carboxylic acids such as acetic acid ($p\underline{K}_a$ ~ 4). In the gas phase, phenol and acetic acid have comparable acidities. Why?

b. In H_2O, acetic acid and benzoic acid have almost the same $p\underline{K}_a$. In the gas phase, benzoic acid is stronger by 8.7 kcal mol^{-1}. Why?

c. A logarithmic plot of gas phase and aqueous phase acidities for substituted benzoic acids is approximately linear. A notable derivation from the plot is p-OH. Its gas phase acidity is anomalously high. Why?

7. Consider the following equilibrium.

$$HOCH_3 + DSCH_3 \rightleftharpoons DOCH_3 + HSCH_3$$

The H-O force constant is about twice as large as the H-S force constant. Which side does the equilibrium favor? You may assume that isotopic substitution on a bond does not alter the force constant, i.e., $k_{HO} = k_{DO}$ and $k_{HS} = k_{DS}$.

8. The ability of a hydroxylic solvent to deprotonate an acid ^+BH depends on whether the hydroxyl group of the solvent bears H or D.

$$ROH + {}^+BH \rightleftharpoons B + R\overset{+}{O}H_2 \qquad K_H$$

$$ROD + {}^+BD \rightleftharpoons B + R\overset{+}{O}D_2 \qquad K_D$$

The value of the ratio K_H/K_D depends on the acidity of the hydroxylic solvent.

Solvent	pK_a	K_H/K_D
2-Chloroethanol	14.7	5.0
2,5-Dinitrophenol	5.2	3.3
Chloroacetic acid	2.7	2.8

Explain the variation in K_H/K_D (ignore any differences between ^+BH and ^+BD).

SOLUTIONS

1. a.

The C-X and C=O dipoles reinforce each other in the equatorial form but cancel to some extent in the axial form. Thus the equatorial form is more polar and should be favored by more polar (higher dielectric constant) solvents. Therefore, 1,4-dioxane has the larger proportion of equatorial-X in each case.

b. Both steric and electrostatic considerations should disfavor the equatorial form, as observed for 2-bromocyclohexanone. The almost parallel orientation of the C-X and C=O dipoles in the equatorial form causes a considerable electrostatic repulsion. This same juxtaposition brings the X and O closer together in the equatorial form. This steric repulsion has been termed the 2-alkyl- (or, here, the 2-halo-) ketone effect, or $\underline{A}^{1,3}$ strain. For X = CH_3, this repulsion is thought to be rather small, but for larger alkyl groups a definite steric effect has been documented. In the equatorial halo series, bromine should have the smallest electrostatic effect but the largest steric effect (although the steric effect is partially compensated by the longer C-Br bond). Both factors contribute to the considerable axial-bromine preference. The situation is less clear with fluorine. The steric effect should be minimized but the electrostatic effect maximized in the equatorial form. The observed equatorial fluorine preference would seem to indicate that the steric effect is dominant, since its magnitude parallels the observed conformer populations for the entire series. The steric effect may also be dependent on the ability of the C-X bond to allow angle deformations. The axial C-Br bond is more easily deformable than the axial C-F bond, so that a larger proportion of axial bromine is permitted. These two steric factors, $\underline{A}^{1,3}$-like strain and ease of angle deformation, probably account for much of the difference between the bromine and fluorine systems. Certainly, the electrostatic repulsion must be secondary, since it predicts the wrong result for fluorine. Attribution of the Br, Cl, F trend entirely to a steric effect, however, would be premature. Difluoroethylene prefers the cis form, in which the fluorine dipoles are electrostatically repulsive. The orbital attraction (see problem 1-9) suggested for difluoroethylene also might apply to the C-F/C=O situation. Thus the equatorial fluorine preference may be due to neither steric nor electrostatic factors but to orbital interactions. The situation requires further clarification. As a footnote, it should be pointed out that the data given in the question were based for the most part on dipole moment experiments. Compared to more modern techniques, dipole moments give very inaccurate conformational equilibrium constants. The trends in the data, on which this question is based, however, should be reliable.

References: E. L. Eliel, N. L. Allinger, S. J. Angyal, and G. A. Morrison, "Conformational Analysis," Wiley-Interscience, New York, NY, 1965, pp 112-14, 460-66; F. Johnson and S. K. Malhotra, J. Am. Chem. Soc., 87, 5492 (1965); N. D. Epiotis, ibid., 95, 3087 (1973); R. C. Bingham, ibid., 98, 535 (1976).

2. a.

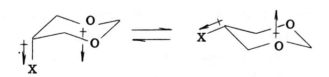

For most solvent effect equilibria, the problem requires determining the relative polarity of the interconverting species. The group dipole of C-O-C bisects the valence angle in the plane of the three atoms. The sum of two such entities gives the illustrated resultant. In the axial form, this resultant is nearly parallel to the C-X dipole. The equatorial form is somewhat less polar because the group dipoles partially cancel. Thus an increase in the dielectric constant of the solvent favors the more polar axial form.

b. Monosubstituted cyclohexanes exhibit their well-known preference for the equatorial position because of unfavorable syn-axial (gauche) H/X interactions in the axial form. The 1,3-dioxane equilibria in solvents of low polarity are dominated by dipolar forces, as described in part (a). In more polar solvents, another factor must become important, because the molecules show a much larger axial preference than monosubstituted cyclohexanes. The 3,5-axial protons of cyclohexane have been replaced by the lone pairs of oxygen in 1,3-dioxane. It appears that oxygen with a lone pair offers less steric interference than carbon with an attached axial proton. With less steric interference in the axial position, the 5 substituents of 1,3-dioxanes exhibit a larger axial/equatorial ratio than analogous substituents in cyclohexane. A purely electronic alternative can be offered to this steric explanation. The F-C-C-O relationship is gauche in the axial conformer, but trans in the equatorial conformer. When X and Y are highly polar, an X-C-C-Y fragment prefers the gauche arrangement, either because of orbital attractions or inhibited delocalization (see problem 1-9). The axial preference for the listed substituents therefore could result from the favorable gauche arrangement. The electrostatic repulsion partially overcomes this steric/orbital effect in solvents of low polarity, with the observed result of higher equatorial preferences. Alkyl substituents also exhibit lower equatorial preferences in polar solvents. As the orbital explanation could not apply to these substituents, a traditional steric effect must be applicable.

44

c. If the effect were entirely steric (i.e., a comparison of axial H in cyclohexanes vs. axial lone pair in 1,3-dioxanes), one might expect the fluorine to be least affected. Its smallest size should render it least sensitive to steric factors. Since $\Delta\Delta\underline{G}^{\circ}$ is largest for fluorine, electrostatic or orbital factors must be more important than steric factors. On the one hand the F–O interaction may be attractive through London forces. Probably more important though is the fact that the high polarity of the C–F bond brings about the strong electronic favoring of the gauche arrangement found in the axial form.

Of the substituents examined, this effect will be largest for F.

References: R. J. Abraham, H.D. Banks, E. L. Eliel, O. Hofer, and M. K. Kaloustian, J. Am. Chem. Soc., 94, 1913 (1972); R. C. Bingham, ibid., 98, 535 (1976).

3. a.

A   B   C

At least three hypotheses can be forwarded.

(i) Steric. In cyclohexyl (A), the axial form is disfavored by two gauche interactions (syn-axial X/H). In the dioxane form B (see problem 3-2), the axial protons have been replaced by the oxygen lone pairs, which offer less steric repulsion, so that the axial form is increased. Similarly, in the exo-methylene system C, one axial proton has been replaced by the π bond, so that steric hindrance should decrease, and there should be more axial form. Furthermore, replacement of CH_2 by $C=CH_2$ flattens the ring slightly, reduces axial-axial interactions, and permits more axial conformer.

(ii) Orbital. The X–C–C–O gauche effect favors the axial form (see problem 1-9). Similarly, there might be an

X-C-C-π gauche effect that favors the axial conformer.

(iii) Electrostatic. Polar interactions between the C-X group and the π cloud would disfavor the axial form and increase the proportion of the equatorial conformer.

b. Since CH_3 is nonpolar, explanations (ii) and (iii) do not apply. The larger proportion of axial CH_3 is readily explained in terms of replacement of one syn-axial CH_3/H steric interaction by a less repulsive CH_3/π interaction, as in (i). For the polar substituents (OCH_3, OH, SCH_3), all of the above explanations must be considered. The observed increase in the equatorial form, however, is inconsistent with the steric (i) and orbital (ii) explanations. The nature of the electrostatic interaction (iii) between the C-X and $C=CH_2$ groups is not clear. The C-X dipole can interact repulsively with either the dipole or the quadrupole of the double bond.

c. Although solvent interactions are probably negligible for CF_2Cl_2, the polar and hydrogen-bonding properties of $CHFCl_2$ exert a profound influence on the equilibrium. The negative end of the C-X dipole is directed into the solvation sphere, and its interactions with solvent (dipolar, hydrogen-bonding) serve to reduce the electrostatic interactions between the axial C-X group and the π electrons. The repulsive interaction described in (b) is thereby reduced, so that the proportion of axial conformer is higher in $CHFCl_2$ than in CF_2Cl_2. The percent axial is even higher for the exo-methylene system (C) in $CHFCl_2$ than for cyclohexyl (A), which is relatively solvent insensitive. Of the explanations offered in (a), we see that the behavior of C in $CHFCl_2$ is consistent with (i) or (ii). Most likely, the reduced intramolecular electrostatic interaction in $CHFCl_2$ permits the steric explanation that applies to CH_3 in all solvents to apply to the polar substituents as well.

d. The acetoxy group is exceptional, in that it is the only polar substituent to exhibit an increase in the axial proportion in the noninteracting solvent CF_2Cl_2. Ester resonance

in fact places positive charge on the innermost atom. In contrast, the oxygen of OCH_3 bears a partial negative charge. The orientation of the OAc group is such that the positive oxygen is directed toward the π bond. This interaction

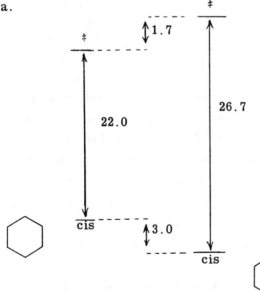

may be attractive and hence raise the axial OAc proportion in comparison to that of OCH_3.

Reference: J. B. Lambert and R. R. Clikeman, J. Am. Chem. Soc., 98, 4203 (1976).

4. a.

The diagram on p 47 shows the relative placement of the cis ground state and the transition state in the two solvents. No data were given on the trans ground state. The enthalpy of transfer of the cis ground state from cyclohexane to cyclohexanone (ΔH° = -3.0 kcal mol^{-1}) and the difference in activation enthalpies ($\Delta\Delta \underline{H}^{\ddagger}$ = 4.7) give the enthalpy of transfer between solvents of the transition state ($\Delta \underline{H}^{t}$ = 1.7). Thus the enthalpy of transfer for the transition state from cyclohexane to cyclohexanone is endothermic.

b. The endothermic $\Delta \underline{H}^{t}$ means that ground state stabilization of the cis form in cyclohexanone ($\Delta \underline{H}^\circ$) is not sufficient to overcome the unfavorable activation energy difference ($\Delta\Delta \underline{H}^{\ddagger}$). The polar cis ground state is stabler in the more polar solvent cyclohexanone than in cyclohexane (the reverse should be the case for the trans ground state, but this information was not given). Similarly, since $\Delta \underline{H}^{t}$ is endothermic, the transition state consequently is the less polar of the two. The two lone pairs in the diradical A are approximately orthogonal

A   B

(90° torsional angle), so that the vector sum of their dipoles is larger than the dipole of a single lone pair. The p-hybridized lone pair in B has no dipole, so that the vector sum of the two lone pair dipoles is about the same as the dipole of a single lone pair. Thus the dipole of A is larger than that of B. Since the endothermic $\Delta \underline{H}^{t}$ indicates that the favored mechanism has the less polar transition state, the isomerization must take place via an inversional process (transition state B).

Reference: P. Haberfield, P. M. Block, and M. S. Lux, J. Am. Chem. Soc., 97, 5804 (1975).

5. (a) $HOAc \rightleftharpoons H^+ + {}^-OAc$

(b) $Py\overset{+}{r}-H \rightleftharpoons Pyr + H^+$

The major difference between the two equilibria is the charge
types. For HOAc (a), two ions (one positive, one negative) must
be solvated on dissociation. Water, with a much higher ε and
the ability to form either the positive or negative end of strong
hydrogen bonds, is much better able to stabilize both charge
types than CH_3OH. Therefore the acidity of HOAc is much lower
in CH_3OH than in H_2O. For $Py\overset{+}{r}$-H (b), the charge type is the
same on both sides of the equilibrium. Therefore the change
from H_2O to CH_3OH has little effect on the acidity constant.
Reference: J. Hine, "Structural Effects on Equilibria in
Organic Chemistry," Wiley-Interscience, New York, NY, 1975,
p 132.

6. a. The negative charge in the acetate ion is appreciably localized
   on the oxygen atoms, whereas the negative charge in the
   phenoxide ion is delocalized over the oxygen and the aromatic ring. In H_2O, the acetate oxygens can be very effectively
   solvated (a "hard-hard" interaction), thereby enhancing the
   acidity. Furthermore, the small methyl group offers little
   steric hindrance to solvation. On the other hand, the larger
   delocalization of charge in the phenoxide ion decreases solvation by H_2O (a "soft-hard" interaction). The larger overall size of phenoxide also causes less effective solvation. In
   the gas phase, the extra solvation of acetate is lost, so that
   the acidities of phenol and acetic acid are comparable. In
   an aprotic solvent such as dimethyl sulfoxide, the special
   solvation of acetate should be much less, so that DMSO acidities should more closely resemble gas phase acidities.

   b. In H_2O, neutral benzoic acid has important contributions
   from charge-separated resonance structures (A). On removal of the proton to form the anion, these contributions
   (B) are much less important because of the double negative

charge. Charge separation enhances solvation, so that neutral benzoic acid is more solvated than neutral acetic acid, in which charge separation is small. To a first approximation, solvation should be similar in the benzoate and acetate anions, in which there is little charge separation. Consequently, the acidity of benzoic acid in H_2O will be lowered in comparison to that of acetic acid. In the gas phase, charge-separated resonance forms are not important because of the absence of stabilizing solvation. The acidity of benzoic acid therefore is increased with respect to that of acetic acid.

c. In H_2O, the more acidic proton on p-hydroxybenzoic acid undoubtedly is on the carboxyl group, because of the generally greater acidity of carboxylic acids over phenols. As pointed out in part (a), however, phenols and carboxylic acids are of comparable acidity in the gas phase. The reason for the deviation from the log-log, H_2O-gas phase plot is that p-hydroxybenzoic acid acts as a carboxylic acid in H_2O but as a phenol in the gas phase. The extra phenolic acidity that generates the large deviation from the plot is probably the result of quinoid resonance (C ⟷ C'), which would be much less important in a carboxylate anion (as in B).

Reference: T. B. McMahon and P. Kebarle, $\underline{J}$. $\underline{Am}$. $\underline{Chem}$. $\underline{Soc}$., 99, 2222 (1977).

7. In isotopic equilibria, the overwhelming factor is the difference in zero point energies between the species on either side. The zero point energy is $\frac{1}{2}h\nu_0$, and the frequency of the vibration (stretching here) is given by

$$\nu_0 = \frac{1}{2\pi} \sqrt{\frac{k}{\mu}} \quad ,$$

in which $\underline{k}$ is the force constant of the vibration and μ is the reduced mass, $\underline{m_1}\underline{m_2}/(\underline{m_1} + \underline{m_2})$, of the system. Any isotopic equilibrium can be solved if all the relevant force constants are known, or if all are known in relation to one. For the present case,

$$\underline{k}_{HO} = \underline{k}_{DO} = 2\underline{k}_{HS} = 2\underline{k}_{DS} = 2\underline{k}.$$

The zero point energies (ZPE) can then be calculated in terms of $\underline{k}$.

$$\text{HOCH}_3 \qquad \frac{h}{4\pi}\left(\frac{2k}{\mu_{HO}}\right)^{\frac{1}{2}} = \frac{h}{4\pi}\left(\frac{2k}{16/17}\right)^{\frac{1}{2}} = \frac{h}{4\pi}(2.13\,\underline{k})^{\frac{1}{2}}$$

$$\text{DSCH}_3 \qquad \frac{h}{4\pi}\left(\frac{k}{\mu_{DS}}\right)^{\frac{1}{2}} = \frac{h}{4\pi}\left(\frac{k}{64/34}\right)^{\frac{1}{2}} = \frac{h}{4\pi}(0.53\,\underline{k})^{\frac{1}{2}}$$

$$\text{DOCH}_3 \qquad \frac{h}{4\pi}\left(\frac{2k}{\mu_{DO}}\right)^{\frac{1}{2}} = \frac{h}{4\pi}\left(\frac{2k}{32/18}\right)^{\frac{1}{2}} = \frac{h}{4\pi}(1.13\,\underline{k})^{\frac{1}{2}}$$

$$\text{HSCH}_3 \qquad \frac{h}{4\pi}\left(\frac{k}{\mu_{HS}}\right)^{\frac{1}{2}} = \frac{h}{4\pi}\left(\frac{k}{32/33}\right)^{\frac{1}{2}} = \frac{h}{4\pi}(1.03\,\underline{k})^{\frac{1}{2}}$$

Thus the ZPE of the left side of the equation is

$$\frac{h\underline{k}^{\frac{1}{2}}}{4\pi} [(2.13)^{\frac{1}{2}} + (0.53)^{\frac{1}{2}}] = \frac{h\underline{k}^{\frac{1}{2}}}{4\pi} (2.19).$$

The ZPE of the right side of the equation is

$$\frac{hk^{\frac{1}{2}}}{4\pi} [(1.13)^{\frac{1}{2}} + (1.03)^{\frac{1}{2}}] = \frac{hk^{\frac{1}{2}}}{4\pi} (2.07).$$

The right side has the lower ZPE and therefore is favored. The actual stretching force constants are $\underline{k}_{H_2O} = 7.76 \times 10^5$ dyne cm^{-1}, $\underline{k}_{D_2O} = 7.94 \times 10^5$, $\underline{k}_{H_2S} = 4.14 \times 10^5$, and $\underline{k}_{D_2S} = 4.46 \times 10^5$. The limitations of the above calculations may be examined by utilizing the more exact values. A general rule for such equilibria has been propounded, that the heavier isotope is attached preferentially to the atom that produces the larger force constant. Thus deuterium is attached preferentially to oxygen $(\underline{k}_{DO} = 2\underline{k}_{DS})$, and the right side (with $DOCH_3$) is favored, as calculated. With this rule and knowledge of relative values of force constants, the direction of any simple isotopic equilibrium can be predicted without calculation.

References: K.B. Wiberg, "Physical Organic Chemistry," Wiley, New York, NY, 1964, pp 273-77; M. Wolfsberg, Acc. Chem. Res., 5, 225 (1972); G. Herzberg, "Molecular Spectra and Molecular Structure II. Infrared and Raman Spectra of Polyatomic Molecules," Van Nostrand, New York, NY, 1945, pp 160-61, 227-31.

8. Since only a ratio of equilibrium constants is given, one must subtract the equilibria as written.

$$ROH + {}^+BH + B + R\overset{+}{O}D_2 \rightleftharpoons B + R\overset{+}{O}H_2 + ROD + {}^+BD$$

The species B is on both sides and subtracts out. Furthermore, ^+BH cancels with ^+BD according to the assumption given. The difference equilibrium therefore reduces to

$$ROH + R\overset{+}{O}D_2 \rightleftharpoons R\overset{+}{O}H_2 + ROD.$$

The constant for this equilibrium, $\underline{K}$, is equal to the ratio, $\underline{K}_H/\underline{K}_D$, which is greater than unity for all solvents. Thus the purely isotopic equilibrium favors the right side, $R\overset{+}{O}H_2 + ROD$. According to the qualitative rule for isotopic equilibria given in problem 3-7, the heavier isotope favors the atom that provides the larger force constant. Thus, since the heavier deuterium favors ROD (right side) over $R\overset{+}{O}D_2$ (left side), $\underline{k}$(ROD) must be greater than $\underline{k}(R\overset{+}{O}D_2)$ ($\underline{k}$ is the O-H or O-D stretching force constant). How is this factor related to the acidity of ROH? As the acidity decreases (larger $p\underline{K}_a$, stronger O-H bond), force constants for the O-H (O-D) bond must get larger. The observed increase in $\underline{K} = \underline{K}_H/\underline{K}_D$ as the acidity decreases means that $\underline{k}$(ROD) is increasing (with decreased acidity) faster than $\underline{k}(ROD_2)$. This is a reasonable result for charged vs. uncharged systems. If the assumption to ignore $^+$BH vs. $^+$BD had not been made, the equilibrium to examine would have been

$$\text{ROH} + {}^{+}\text{BH} + R\overset{+}{O}D_2 \rightleftharpoons R\overset{+}{O}H_2 + \text{ROD} + {}^{+}\text{BD}.$$

Neglect of $^+$BH vs. $^+$BD is equivalent to assuming that the $^+$B-H force constant is smaller than the O-H. For neutral bases $(\underline{k}^+{}_{BH} < \underline{k}_{OH})$ and for most carbon bases $(\underline{k}_{CH} < \underline{k}_{OH}$ normally), this is probably a good assumption.

Reference: M. Wolfsberg, Acc. Chem. Res., $\underset{\sim}{5}$, **225** (1972).

ALSO SEE PROBLEMS 4-2, 4-6, 4-7, 4-8, 5-2, 5-4. 5-5.

4
EQUILIBRIUM ELECTRONIC EFFECTS

PROBLEMS

1. The acidity of 9,10-dihydroanthracene-1-carboxylic acids (A) and of 9,10-ethanoanthracene-1-carboxylic acids (B) is dependent on the nature of the 8 substituent (X).

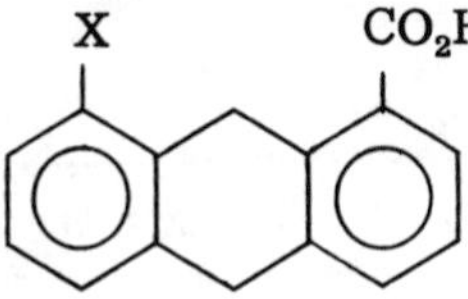

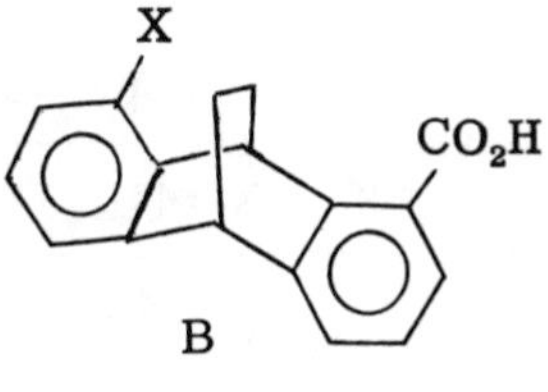

X	pK_a (A)	pK_a (B)
H	5.56	5.99
OCH_3	--	6.15
Cl	5.71	6.24
CO_2^-	6.31	6.89

a. Explain why the pK_a increases on going down the columns (increased electronegativity).

b. Why are the systems A generally more acidic (lower pK_a) than the systems B?

2. <u>Related subject</u>: <u>equilibrium solvent effects</u>

The pK_a's in 50% aqueous ethanol of methylammonium salts ($X = \overset{+}{N}$) have been compared with those of fully methylated carbon systems ($X = C$) in acyclic (A), monocyclic (B), and bicyclic (C) compounds.

A $(CH_3)_3X-(CH_2)_n CO_2H$ $\underline{n} = 1\ (\alpha),\ 2\ (\beta),\ \text{or}\ 3\ (\gamma)$

B

$CO_2H\ \alpha,\ \beta,\ \text{or}\ \gamma\ \text{to}\ X$

C

$CO_2H\ \alpha,\ \beta,\ \text{or}\ \gamma\ \text{to}\ X$

Series	$pK_a\ (\alpha)$		$pK_a\ (\beta)$		$pK_a\ (\gamma)$	
	$X = \overset{+}{N}$	$X = C$	$X = \overset{+}{N}$	$X = C$	$X = \overset{+}{N}$	$X = C$
A	2.93	6.37	4.16	6.12	4.96	6.13
B	2.96	6.71	4.22	6.30	4.69	6.26
C	2.97	6.86	4.35	6.61	4.57	6.60

a. Polar effects should be minimal in the carbon compounds. Account for why the pK_a's for the α molecules are slightly larger than those for the β and γ molecules in all three series (A, B, C for $X = C$).

b. After correction for a steric effect, the pK_a's for the ammonium compounds ($X = \overset{+}{N}$) give a linear plot against the reciprocal distance $1/r$ between N^+ and the dissociable proton for all three series (A, B, C). How might the correction for a steric effect be applied? Is the linear plot consonant with the inductive or the field model for the σ polar effect (see problem 4-1)?

c. The $\underline{K}_a$'s for the Y compounds in H_2O are 1.10×10^{-4} for A, 2.40×10^{-4} for B, and 4.79×10^{-4} for C. Are these data consonant with the inductive or the field model?

3. Consider the following Hammett substituent constants.

Substituent	σ_m	σ_p
OCH_3	0.12	-0.26
SCH_3	0.14	0.01
OCF_3	0.37	0.35
SCF_3	0.37	0.42

Explain the effects of S and F substitution for O and H, respectively.

4. Related subject: conformational analysis

a. Which series of σ constants should be used to calculate ρ for the Hammett plot of the following equilibrium? Why?

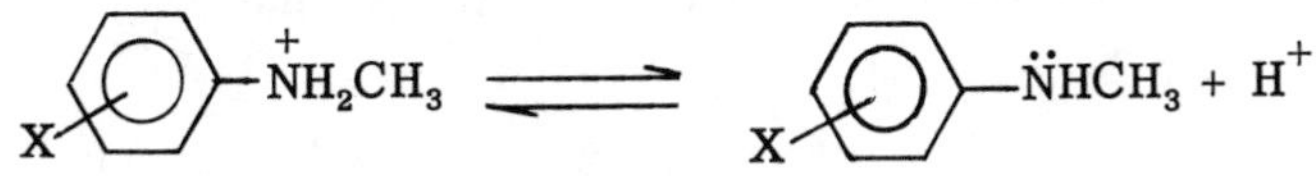

b. Which series of σ constants is appropriate for the following equilibrium? Why? Compare this situation with that in (a).

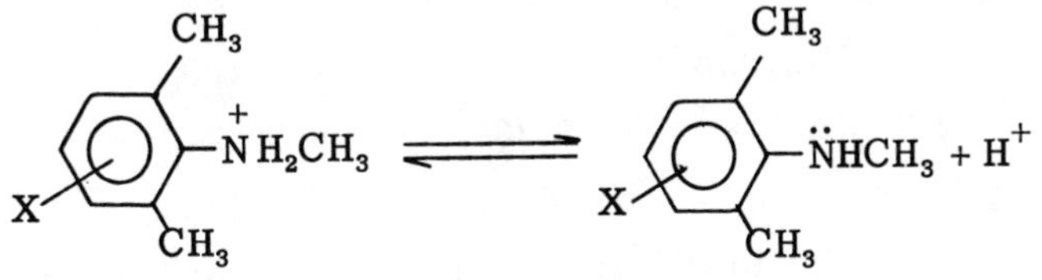

5. Related subject: conformational analysis

a. What is the sign and approximate magnitude of ρ for the following equilibrium? Explain.

Electronic Effects on Equilibria

$$Ar_3C^+ + H_2O \xrightleftharpoons{H_2SO_4} Ar_3C\text{-}OH + H^+$$

b. Should the ρ for the following equilibrium be larger or smaller than that in part (a)? Explain.

$$Ar_2\overset{+}{C}H + H_2O \xrightleftharpoons{H_2SO_4} Ar_2CH\text{-}OH + H^+$$

6. Related subject: equilibrium solvent effects

a. Anilines and formic acid are in equilibrium with formanilides. What is the sign of ρ for this equilibrium as written? Explain.

$$Ar\text{-}\overset{..}{N}H_2 + HCO_2H \rightleftharpoons Ar\text{-}\overset{..}{N}H\text{-}\overset{\overset{\displaystyle O}{\|}}{C}\text{-}H + H_2O$$

b. The ρ for the ionization of anilinium ions in H_2O is large and positive. As the solvent is changed from pure H_2O to more

$$Ar\text{-}\overset{+}{N}H_3 + H_2O \rightleftharpoons Ar\text{-}\overset{..}{N}H_2 + H_3O^+$$

and more 1,4-dioxane, ρ increases (2.77 in pure H_2O, 3.26 in 20% dioxane, 3.57 in 70% dioxane). Explain why in terms of solvent properties and the sources of substituent effects.

7. Related subjects: organothermochemistry, equilibrium solvent effects

Consider the following thermodynamic data for the ionization of benzoic acids in H_2O at 25 °C (in cal mol^{-1}).

Acid	ΔG°	ΔH°	ΔS° (eu)
p-Toluic	5962 ± 1	−134 ± 9	−20.44 ± 0.03
m-Toluic	5800	−91	−19.75
Benzoic	5732	−67	−19.44
m-Methoxybenzoic	5583	22	−18.65
p-Chlorobenzoic	5437	100	−17.90
m-Nitrobenzoic	4720	421	−14.41
p-Nitrobenzoic	4671	432	−14.21

a. In order for the free energy of a given equilibrium to be proportional to that of the benzoic acid series ($\Delta\Delta G^\circ \propto \Delta\Delta G^{\circ\prime}$), one of three conditions must hold. Both series must be isoenthalpic ($\Delta\Delta H^\circ = 0$), both series must be isoentropic ($\Delta\Delta S^\circ = 0$), or the enthalpy changes for both series

must be proportional to the respective entropy changes
($\Delta\Delta\underline{H}^\circ = \beta\Delta\Delta\underline{S}^\circ$). According to the above data, to which category does the ionization of benzoic acid belong (and hence all equilibria that follow the Hammett equation)?

b. Substituent effects are generally described by phrases such as "p-nitrobenzoic acid is more acidic than m-nitrobenzoic acid because the nitro group withdraws electrons by resonance very strongly in the para position, but less so in the meta, and because polar electron withdrawal is still important in the para position despite the greater distance than the meta position from the charged site." Discuss the source of resonance and polar effects in light of the results in part (a).

8. Related subjects: conformational analysis, equilibrium solvent effects

a. The $\underline{K}_a$ (solvent H_2SO_4) for removal of the first proton in 1,10-phenanthroline-2H$^+$ (A) is 35.5, but that for 2,2'-dipyridine-2H$^+$ (B) is 3.3. Up to four reasons can be suggested for this 11-fold difference. Give as many as you can.

A   B

b. The $\underline{K}_a$ for removal of the proton in the N-methyl analogue of A is 129, for the N-methyl analogue of B, 0.49. Why does $\underline{K}_a$ go up for A (ratio 3.6) but down for B (ratio 0.15)? This answer should utilize your responses to (a) and demonstrate which are not correct.

A-CH$_3$ B-CH$_3$

SOLUTIONS

1. a. The 9,10-dihydro- and 9,10-ethanoanthracenes isolate the ionizing CO_2H group from the perturbing X group by saturated carbon atoms. Consequently, resonance and π inductive effects should be negligible. The normal σ inductive effect (through-bond charge polarization) of electron-withdrawing groups like Cl or OCH_3 is acid strengthening (compare CH_3CO_2H and CCl_3CO_2H). The opposite result is observed in the present systems. The field effect (through-space charge polarization) is angle dependent and can produce either acid strengthening or acid weakening, depending on the relative orientation of the substituent dipole with the carboxylate negative charge. The structure of A and B constrains the negative end of the C-X dipole toward the carboxylate group. This orientation is clearly acid weakening, as observed.

Thus the field effect offers the only plausible explanation for these reversed substituent effects. The field effect might also supply the dominant contribution to the normal acid-strengthening effect observed when electron-withdrawing groups are structurally unconstrained and can orient their dipoles away from the carboxylate anion (as in CCl_3CO_2H). A complete separation of inductive and field effects, however, can never be made because the phenomena are interdependent. The CO_2^- as a substituent is slightly electron-withdrawing ($\sigma_p = +0.13$, compared to $+0.23$ for Cl). Its extremely large acid-weakening effect in A and B is due to the stronger monopole-monopole interaction (CO_2^-/CO_2^-) than dipole-monopole (e.g., Cl/CO_2^-).

b. System A is nearly planar, but the groups (X and CO_2H) are probably too far away for a direct or solvent-mediated steric effect. The bridge in B pulls the two aromatic rings closer together. As a result, the C-X dipole and the CO_2^- monopole are closer together. The larger field effect then makes the X groups even more acid weakening in B than in A, so the systems A are more acidic for a given X group. Steric crowding in B, with anion desolvation, would also be acid weakening, but distances (about $4.5 Å$) are probably too large.

References: R. Golden and L. M. Stock, J. Am. Chem. Soc., 94, 3080 (1972); L. P. Hammett, "Physical Organic Chemistry," 2nd ed, McGraw-Hill, New York, NY, 1973, pp 374-76.

2. a. Resonance effects should be small in these systems because of their saturated nature. Furthermore, polar effects should not be significant in the carbon series. The higher pK_a's for the α compounds can be attributed to steric hindrance to solvation. The closeness of the carboxyl group to the quaternary carbon inhibits effective solvation of the anion and reduces acidity. The β and γ values are essentially equal for each series (A: 6.12, 6.13; B: 6.30, 6.26; C: 6.61, 6.60), as the steric effect should no longer be important at these distances.

b. The β system can be used as the model for no steric effect in the carbon series, so the A-α pK_a should be reduced by 0.25, the B-α by 0.41, and the C-α by 0.25 pK_a unit. The acidities in the ammonium series should be dependent on these same steric effects and in addition on the polar effect of $\overset{+}{N}$, but not on resonance effects. If the steric corrections derived in the carbon series can be carried over to the ammonium series, the pK_a's for the α compounds become 2.68 (A), 2.55 (B), and 2.72 (C). These reduced values are determined entirely by the polar effect. The linear plots are consonant with a field mechanism, which varies inversely with the distance, according to the Bjerrum formulation. An angular dependence is also present but does not alter the plot of the radial dependence alone.

c. According to the inductive model, the polar effect depends on the number of bonds between X and CO_2H and on the number of transmitting pathways. In the γ series, the number of bonds is constant. The ratios (A/B/C) of transmitting pathways are 1/2/3, and the observed ratios are 1/2.18/4.35. The acidities of the monocyclic compound (B) to some extent and the bicyclic compound (C) to a large extent are much greater than predicted by the inductive model. Thus the acidities are consistent with the field model (part (b)) and are inconsistent with the inductive model. The authors emphasize that a distinction between the two transmission models thus is not necessary, the inductive model being a special case in which the dielectric in the field model is described atomistically.

Reference: C. A. Grob, A. Kaiser, and T. Schweizer, <u>Helv.</u> <u>Chim</u>. <u>Acta,</u> <u>60</u>, 391 (1977).

3. According to the Swain-Lupton formalism, σ_m is about 22% resonance and σ_p about 53% resonance. Thus one can discuss σ_m as primarily a polar substituent constant, but σ_p has contributions from both polar and resonance phenomena. The values for OCH_3 are the oft-quoted result of the balance between its polar electron-withdrawing and its resonance electron-donating character. The value of σ_m = 0.12 represents a modest electron-withdrawing effect, associated with the electronegativity of oxygen. The resonance donation from the para position (C$\overset{\cdot\cdot}{\underset{}{-}}\ddot{O}CH_3$ $\longleftrightarrow$ C$=\overset{+}{O}CH_3$), however, overcomes the polar character and results in a net electron-donating σ_p (-0.26). The SCH_3 substituent has about the same σ_m as OCH_3, but a considerably less negative σ_p. One concludes that the polar effect of SCH_3 must be about the same as that of OCH_3, but the resonance effect is considerably reduced. Although the electronegativity of sulfur is not so large as that for oxygen (current literature values of the S electronegativity, however, are almost certainly too small), other factors such as its much higher polarizability may be com-

pensatory, so that the net polar effects of OCH_3 and SCH_3 (and hence their values of σ_m) are similar. Resonance interaction of the OCH_3 group with the phenyl group involves 2p-2p overlap, whereas that of the SCH_3 group requires 3p-2p overlap ($C-\overset{\cdot\cdot}{S}CH_3$ $\longleftrightarrow C=\overset{+}{S}CH_3$). The 3p orbitals are larger than the 2p, so that 3p-2p overlap is less effective than 2p-2p overlap. With a smaller resonance effect, the SCH_3 group in the para position is less able than OCH_3 to overcome the opposite polar effect, so that σ_p for SCH_3 becomes less negative (more positive) than that for OCH_3. The σ_m for OCF_3 is larger than that of OCH_3 because the electron-withdrawing fluorine atoms make the group more polar. By the same token, the fluorine atoms lessen the resonance interaction in the para position ($C-\overset{\cdot\cdot}{O}CF_3 \longleftrightarrow C=\overset{+}{O}CF_3$) and increase σ_p, since the oxonium positive charge is destabilized by the electron-withdrawing fluorine atoms. Because σ_m and σ_p are about the same magnitude (0.37, 0.35), the resonance effect must be almost entirely quenched. The equality of σ_m for OCF_3 and SCF_3 again indicates that the polar effects of sulfur and oxygen are about the same. The fact that σ_p for SCF_3 is larger than σ_m means that the group is more electron withdrawing in the para than in the meta position. If the polar effect is smaller in the para position (the positive end of the substituent dipole is further from the CO_2H group in the defining benzoic acid equilibrium), the increased σ_p must be the result of some type of resonance phenomenon. One possibility is $\pi-\sigma^*$ overlap, using unoccupied antibonding orbitals on fluorine. For the effect to be larger for SCF_3 than for OCH_3, one must invoke the lower bond energies of C—S bonds compared to C—O bonds.

Reference: O. Exner, Collect. Czech. Chem. Commun., 31, 65 (1966).

4. a. Direct conjugation is possible between certain substituents X at the para position and the amine site of the free base on the right side of the equilibrium in (a). The situation is analogous to the acid dissociation of phenols. For a regular

σ_p constant to be used, the proportional mix of resonance, polar, and steric effects must be constant. Because direct conjugation of the oxide ion in phenoxide can occur with certain substituents and because direct conjugation is impossible between the carboxylate group and these substituents in benzoic acids, the proportion of resonance interaction is higher for phenoxide than for benzoate. The σ_p constant defined by the benzoic acid ionization therefore is inappropriate, and another constant (σ_p^-) with a higher resonance proportion must be used for substituents capable of direct conjugation with the negative charge. The Swain-Lupton approach gives 53% resonance for σ_p^- and 56% for σ_p^-. All meta substituents $(\sigma_m = \sigma_m^-)$ and nonconjugating para substituents (CH_3, OCH_3) do not require a new σ_p^- value. The amine site of the aniline can conjugate with these electron-withdrawing substituents in the same manner as the oxide in the phenoxide ion (see above resonance structures). This higher implied proportion of resonance means that the benzoic acid σ_p constant again is inappropriate. The reasonable analogy between the amine and the phenoxide suggests that σ_p^- should be used for conjugating para substituents.

b. In the 2, 6-dimethylanilines, the site of protonation can no longer conjugate directly with para substituents. The ortho methyl groups have a strong steric interaction with the

N-methyl group. A planar arrangement is not possible, so that the $NHCH_3$ group rotates out of conjugation with the phenyl ring. In this arrangement, a para substituent is not

Planar, conjugation   Nonplanar, no conjugation

able to conjugate directly with the lone pair, as is also the case with the carboxylate group in the benzoic acid series. Thus a σ_p constant is more appropriate than a σ_p^- constant in systems with sterically inhibited resonance. In this series, meta substituents should be omitted, since σ_m would be severely perturbed by the ortho methyl groups. It should be recognized that a continuum of steric situations can exist, ranging from no conjugation to full conjugation. Rather than employ a large number of different σ constants that would be appropriate to any given situation, it is normal to choose either σ, σ^-, or a linear combination such as $\sigma_p + \underline{r}(\sigma_p^- - \sigma_p)$. This approach is due to Yukawa and Tsuno and has probably been used more frequently in cases of through conjugation of positive charge, $\sigma_p + \underline{r}(\sigma_p^+ - \sigma_p)$. The Hammett ρ is obtained from the meta substituents, and then the value of $\underline{r}$ (a measure of the extent of conjugation) is obtained empirically to bring the conjugating para substituents onto the same line.

Electronic Effects on Equilibria

Reference: R. D. Gilliom, "Introduction to Physical Organic
Chemistry," Addison-Wesley, Reading, MA, 1970, pp 150-52,
156.

5. a. The defining equilibrium, ionization of benzoic acid in H_2O
    at 25 °C, has a ρ of +1.00. The left side of the benzoic acid
    equilibrium is neutral and the aryl moiety on the right is
    negatively charged. The ρ is positive because an electron-
    withdrawing group stabilizes the right side and increases the
    equilibrium constant. The equilibrium in the present question
    has a positively charged moiety on the left and a neutral on
    the right. Thus the change in charge is the same (+1 $\longrightarrow$
    0) as in the benzoic acid equilibrium (0 $\longrightarrow$ -1), a net
    change of -1, and the sign of ρ must be the same for the two
    equilibria. The site of charge density in Ar_3C^+ is one atom
    closer than in $ArCO_2^-$, so the magnitude of ρ should be larger
    in the former case. In fact ρ is 6.67 for the given equili-
    brium (Ar = C_6H_5).

    b. The ρ for the diaryl system should be larger than that in the
    triaryl system. The two phenyl rings can conjugate strongly
    with the cationic center, since steric interference between
    ortho protons can be relieved to a certain extent by angle
    bending toward the H. In the triaryl system, however, angle

bending would simply aggravate one ortho/ortho interaction in
order to improve another. The only way to reduce these interactions is by out-of-plane rotation. In a propeller-like conformation, three ortho protons are up and three are down.
The nonbonded interactions are relieved at the expense of
aryl-C^+ conjugation, since the p orbitals are no longer

66

optimally aligned for overlap. Because of the reduced conjugation, the substituent effect in the triaryl system is smaller, and ρ is also smaller (3.55 vs. 6.67 for Ar = C_6H_5).

References: H. H. Jaffé, Chem. Rev., 53, 191 (1953); K. B. Wiberg, "Physical Organic Chemistry," Wiley, New York, NY, 1964, p 290. The author thanks Professor A. Streitwieser, Jr., for suggesting this problem.

6. a. This is an equilibrium problem. Students frequently offer a discussion of mechanisms and transition states, which are entirely irrelevant. One need only examine the importance of substituent effects on both sides of the equation. Consider the effect of a p-nitro group, which shifts the benzoic acid equilibrium (with a positive ρ) to the right. The left-hand aromatic species (the aniline) in the equilibrium in question is stabilized by through conjugation (A). The right-hand

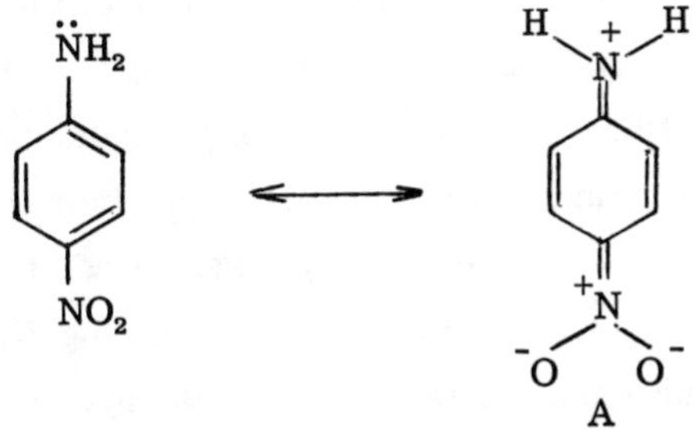

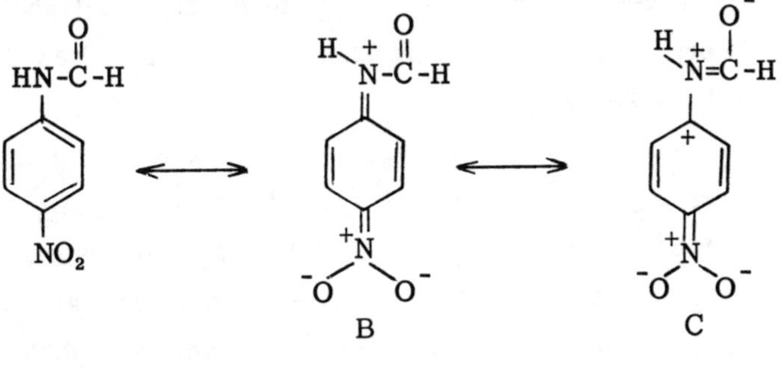

aromatic species (the formanilide) also appears to be stabilized by p-nitro substitution. The stabilization represented by B, however, is reduced by interaction of the nitrogen lone pair with the carbonyl group (C, amide resonance). The formal charge on the amine nitrogen is close to the positive end of the carbonyl dipole in B(C=O $\longleftrightarrow$ $\overset{+}{C}$-$\overset{-}{O}$) or the formal positive charge in the ring in C. Thus nitro stabilizes the right side of the equilibrium more than the left side, in contrast to benzoic acid ionization, so the sign of ρ is opposite, or negative (-1.43).

b. Addition of 1,4-dioxane to the solvent medium decreases its dielectric constant (ε). As interactions with the solvent lessen (because of its lower polarity), intramolecular interactions such as substituent effects increase. For example, the field effect, according to the Kirkwood-Westheimer approach, is inversely proportional to the dielectric constant ($\underline{V} \propto 1/\varepsilon$). Thus a decreased dielectric constant enhances the field effect. Increased sensitivity of the reaction to substituent effects results in an increased ρ. Alternatively, the variation of ρ with added dioxane can be attributed to changes in the amount of hydrogen bonding between aniline and the solvent. In pure H_2O, hydrogen bonds between the solvent and the nitrogen lone pair reduce ring-NH_2 conjugation. Introduction of dioxane reduces the hydrogen bonding, increases conjugation, and hence enhances the sensitivity of the equilibrium to substituent effects as measured by ρ.

References: H. H. Jaffé, Chem. Rev., 53, 191 (1953); K. B. Wiberg, "Physical Organic Chemistry," Wiley, New York, NY, 1964, pp 289-90.

7. a. Recall that $\Delta \underline{G}^\circ = \Delta \underline{H}^\circ - \underline{T} \Delta \underline{S}^\circ$. Thus the Hammett linear free energy relationship $\Delta \Delta \underline{G}^\circ \propto \Delta \Delta \underline{G}^{\circ \prime}$ requires (i) $\Delta \Delta \underline{H}^\circ = 0$, (ii) $\Delta \Delta \underline{S}^\circ = 0$, or (iii) $\Delta \Delta \underline{H}^\circ = \beta \Delta \Delta \underline{S}^\circ$. The $\Delta \Delta \underline{X}^\circ$ corresponds to $\Delta \underline{X}^\circ_{\underline{Y}} - \Delta \underline{X}^\circ_{\underline{H}}$, with $\Delta \underline{X}^\circ_{\underline{Y}}$ referring to any substituent Y and $\Delta \underline{X}^\circ_{\underline{H}}$ referring to no substituent. The benzoic acid equilibrium is clearly neither isoenthalpic nor isoentropic, since $\Delta \underline{H}^\circ$ varies from 4671 to 5962 cal mol^{-1} and $\Delta \underline{S}^\circ$ varies from -20.44 to

to -14.21 eu. A plot of $\Delta \underline{H}^{\circ}$ vs. $\Delta \underline{S}^{\circ}$ (or of $\Delta \underline{G}^{\circ}$ vs. either $\Delta \underline{H}^{\circ}$ or $\Delta \underline{S}^{\circ}$), however, is linear with a correlation coefficient (eleven points altogether--four points in addition to those given above) of 0.9992. Thus Hammett systems should fall into category (iii) above and exhibit linear $\Delta \underline{H}^{\circ}$ vs. $\Delta \underline{S}^{\circ}$ plots. The equation $\Delta\Delta \underline{H}^{\circ} = \beta \Delta\Delta \underline{S}^{\circ}$ is called the isoequilibrium relationship, and β is the (isoequilibrium) temperature at which substituent effects disappear, i.e., $\Delta\Delta \underline{G}^{\circ} = 0$. A system is isoenthalpic if $\beta = 0$, isoentropic if $\beta = \infty$.

b. From the above data, substituent effects must be composed of both enthalpic and entropic factors, and these factors must be proportional. Polar effects (inductive and field) would appear to be basically enthalpic phenomena, since coulombic interactions of charges are involved. Resonance donation or removal of charge also involves a considerable coulombic component, since the end result is frequently the better or worse arrangement of formal charges. Entropic contributions can come from several sources. Thus resonance interaction with the aromatic ring reduces rotational degrees of freedom by hindering rotation about the X-Ar bond. Charge concentration

at certain centers on the aromatic ring by either resonance or induction requires solvent organization. (One entropy source that does not exist but that many students suggest is from mixing of resonance structures; since resonance structures have no independent existence but are multiple representations of a static situation, such mixing is nonexistent.) Both the solvent effect and the rotational effect become larger as the resonance and polar effects become stronger. For this reason the entropy change is proportional to the enthalpy change, and the free energy change is proportional to both factors.

References: J. E. Leffler and E. Grunwald, "Rates and Equilibria
of Organic Reactions," Wiley, New York, NY, 1963, pp 324-26,
341-42; P. D. Bolton, K. A. Fleming, and F. M. Hall, J. Am.
Chem. Soc., 94, 1033 (1972).

8. a. (i) Steric repulsion between the acidic protons, $^+$N-H)(H-N$^+$,
           is larger in A, since B can twist about the central bond
           and move the protons apart. Since there is greater steric
           repulsion in A, the system releases its proton more
           easily.

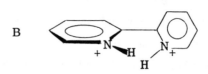

   (ii) Deprotonation of A gives a hydrogen bond, $^+$N—H⸽⸽⸽:N,
          that is stronger than that formed in B, because the atomic
          distances are more favorable in A. The distance is
          slightly too long in B.

   (iii) Coulombic repulsion between the two positive charges is
           stronger in A, since the bond twist in B provides some
           relief.

   (iv) The positive charges are more fully delocalized in the
          planar A, which therefore is less well solvated in H_2SO_4
          than B. Removal of the first proton decreases the differ-
          ential between A and B, giving a net stabilization for A.
       None of these explanations can be eliminated on the basis of
       the data in (a) alone.

   b. Reasons (iii) and (iv) should be about constant for the RH_2^{2+}
       and $RHCH_3^{2+}$ systems and therefore must not be applicable.
       The 3.6-fold increase in the ionization of the phenanthroline
       A suggests that the steric effect (i) is important. The
       methylated A system clearly has greater steric congestion
       than the protonated A system, and deprotonation will provide
       greater relief for the CH_3 than the H system. The ratio for

the ethyl system, $\underline{K}(RHC_2H_5^{2+})/\underline{K}(RH_2^{2+})$, is even larger, 26, since the steric effect is worse. It should be noted that the statistical factor for the ratio $\underline{K}(RHCH_3^{2+})/\underline{K}(RH_2^{2+})$ is $\frac{1}{2}$, since the H system has two ionizable protons but the CH_3 system has only one. Thus the ratio in the absence of other factors should be 0.5. In the bipyridyl system B, replacement of one H by CH_3 must have a negligible steric effect, since the ratio $\underline{K}(RHCH_3^{2+})/\underline{K}(RH_2^{2+})$ is less than unity, i.e., ionization is less likely in the methylated form. In B, no hydrogen bond is possible when the H is replaced by CH_3, since there is no remaining proton to form the bond. Thus in the absence of a steric effect (or reduced importance thereof),

the presence or absence of a hydrogen bond has some effect (reduction from the statistical 0.5 to 0.15). The steric effect in A actually must overcome this lack of a hydrogen bond, so the true measure of the steric effect is the ratio of 3.6 (for A) to 0.15 (for B), or 24. The corresponding ratio for ethyl is 170. It therefore appears that the difference between the fully protonated A and B systems (a), 35.5 and 3.3, is due both to a steric effect (i) (since it is increased by replacement of H by CH_3) and to hydrogen bonding (ii) (since hydrogen bonding is important even in the less likely system B). To the extent that hydrogen bonding is stronger in A than B, the steric effects of 24 for CH_3 and 170 for C_2H_5 would have to be increased.

Reference: O. T. Benfey and J. W. Mills, <u>J. Am. Chem. Soc.</u>, <u>93</u>, 922 (1971).

ALSO SEE PROBLEM 3-6.

IONIC EQUILIBRIA

PROBLEMS

1. a. Although the organic chemist makes frequent use of the acid
dissociation constant ($p\underline{K}_a$), the base hydrolysis constant
($p\underline{K}_b$, for the equilibrium given below) is less

$$X^- + H_2O \rightleftharpoons HX + {}^-OH$$

frequently encountered. Derive a simple expression relating
the $p\underline{K}_b$ for X^- with the $p\underline{K}_a$ for its conjugate acid, HX.

   b. 2,4,6-Trinitroaniline ($Ar-NH_2$) is 7.1% protonated in 90%
H_2SO_4, as measured colorimetrically. What is the $p\underline{K}_a$ of the
conjugate acid ($Ar-\overset{+}{N}H_3$)? Use a table of acidity functions.

   c. 2,4-Dichloro-6-nitroaniline (A) and 2,3,6-trichlorobenzamide
(B) in their protonated forms both have a $p\underline{K}_a$ in H_2SO_4/H_2O
of about -3.30. For each case separately, calculate what
percent H_2SO_4 is necessary for there to be about 1% of the

73

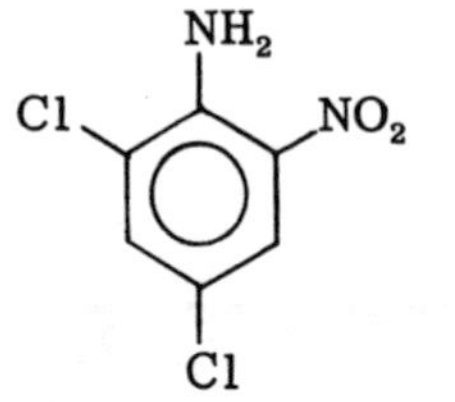

A 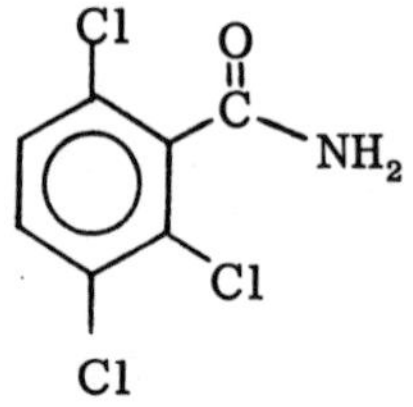   B

base present. Use a table of acidity functions.

2. <u>Related subject</u>: <u>equilibrium solvent effects</u>

One reason given for the strong dependence of $\underline{H}_0$-type functions on the nature of the indicator ($\underline{H}_0$ for Ar-NH$_2$, $\underline{H}'''$ for Ar$_2$NR, $\underline{H}'_R$ for Ar$_2$C=CH$_2$, etc.) is that there is great variability in the number and strength of hydrogen bonds to H$_2$O in the conjugate acids.

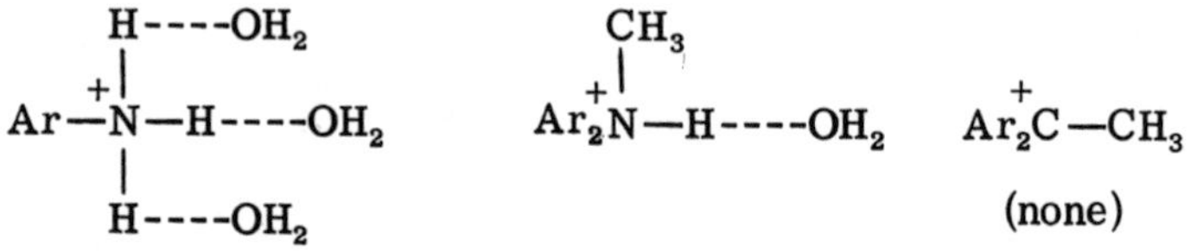

a. Rationalize such an explanation.

b. The acidity function $\underline{H}_B$ for ketones (based on benzophenones) is very similar to $\underline{H}_0$. Using the hydrogen-bonding hypothesis, rationalize this situation.

c. On the basis of this hypothesis, should the $\underline{H}_-$ functions for the following types of indicators be very different?

        Ar$-$NH$_2$ Ar$_2$NH   (CN)$_2$CH$_2$

3. a. Derive a mathematical relationship between $\underline{H}_{OH}$ and $\underline{H}_R$ for the species (C$_6$H$_5$)$_2$CHOH. The defining equilibria are

$$R-\overset{+}{O}H_2 \rightleftharpoons R-OH + H^+ \qquad\qquad \text{for } \underline{H}_{OH}$$

$$\text{and } R^+ + H_2O \rightleftharpoons R-OH + H^+ \qquad\qquad \text{for } \underline{H}_R.$$

b. The function $\underline{H}_{OH}$ + $\log a_{H_2O}$ does not show the same behavior as a function of acid concentration as $\underline{H}_R$. A plot of $\underline{H}_R$ vs. percent H_2SO_4 is more negative than that of $\underline{H}_{OH}$ + $\log a_{H_2O}$. Suggest an explanation.

4. Related subject: equilibrium solvent effects

Bunnett and Olsen examined the problem of the dependence of the acidity function on the structure of the base through assessment of the solvation factor. They found that a plot of $\log(\gamma_{XH^+}/\gamma_{X}\gamma_{H^+})$ vs. $\log(\gamma_{InH^+}/\gamma_{In}\gamma_{H^+})$ is approximately linear with a slope of $(1 - \phi)$ and a zero intercept. In this expression, X is any base

$$\log \frac{\gamma_{XH^+}}{\gamma_X \gamma_{H^+}} = (1 - \phi) \log \frac{\gamma_{InH^+}}{\gamma_{In} \gamma_{H^+}}$$

and In is a Hammett base (a primary aromatic amine). The quantity ϕ has been called the solvation parameter. The plot is carried out by systematically varying the acidity of the solvent, e.g., from 10 to 50% H_2SO_4.

a. Recast this expression in terms of ϕ, $\log [H^+]$, $\underline{H}_0$, and $\underline{H}_X$ only. The parameter $\underline{H}_X$ is the acidity scale for any series of similarly constituted bases, e.g., $\underline{H}_I$ for indoles.

b. What is ϕ when X is a Hammett base, In?

c. What is ϕ when X is H_2O?

d. What, in terms of activity coefficients, does ϕ appear to measure? Examine the following data in order to make your assessment.

$\underline{H}_X$	ϕ (approximately)
$\underline{H}_R$ (see problem 5-3)	-1.3
$\underline{H}_I$ (indoles)	-0.5
$\underline{H}'''$ (see problem 5-2)	-0.4
$\underline{H}_{RSR}$ (thioethers)	-0.3
$\underline{H}_A$ (see problem 5-1)	0.4
$\underline{H}_{ROR}$ (ethers)	0.8

5. Related subject: equilibrium solvent effects

a. The pK_a can be calculated for the protonated form ^+InH of a Hammett base from the expression $H_0 - \log([In]/[^+InH])$. If the species is not a Hammett base, the pK_a can still be calculated for ^+XH from the Bunnett ϕ for the X base, even if the acidity function H_X is not available for the conditions of the single measurement. Derive an expression for the pK_a of ^+XH in terms of $[X]/[^+XH]$, H_0, ϕ, and $[H^+]$. See problem 5-4 for the ϕ equation.

b. If ϕ is not known for a given base, the pK_a can still be obtained from a series of measurements in which the acid concentration is changed, rather than by the single measurement used in (a). What quantities must be plotted against each other to give ϕ from the slope and the pK_a from the intercept for any ^+XH?

c. This procedure (b) was used to give the following pK_a data. What is the significance of the variation of ϕ in this series of structurally similar ketones?

Base(X)	$pK_a(^+XH)$	ϕ
$CH_3-\overset{O}{\overset{\|}{C}}-CH_3$	-2.85	+0.75
$CH_3-\overset{O}{\overset{\|}{C}}-\triangleleft$	-3.27	+0.55
$CH_3-\overset{O}{\overset{\|}{C}}-C_6H_5$	-4.36	+0.40
$C_6H_5-\overset{O}{\overset{\|}{C}}-p\text{-}CH_3O-C_6H_4$	-4.18	+0.26
$C_6H_5-\overset{O}{\overset{\|}{C}}-2,4,6\text{-}(CH_3O)_3-C_6H_2$	-3.59	-0.11

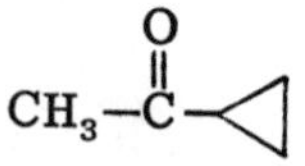

SOLUTIONS

1. a. For clarity, the concentration of water, $[H_2O]$, will be retained in the following expressions. Normally, $[H_2O]$ changes
   so little that it is incorporated into the $\underline{K}_a$ or $\underline{K}_b$. For the
   base hydrolysis,

   $$X^- + H_2O \;\rightleftharpoons\; HX + {}^-OH$$

   $$p\underline{K}_b = -\log \frac{[HX][{}^-OH]}{[X^-][H_2O]} \;.$$

   For the acid dissociation,

   $$HX + H_2O \;\rightleftharpoons\; H_3O^+ + X^-$$

   $$p\underline{K}_a = -\log \frac{[H_3O^+][X^-]}{[HX][H_2O]} \;.$$

   $$p\underline{K}_a + p\underline{K}_b = -\log \frac{[H_3O^+][X^-][HX][{}^-OH]}{[HX][H_2O][X^-][H_2O]}$$

   $$= -\log \frac{[H_3O^+][{}^-OH]}{[H_2O]^2}$$

   $p\underline{K}_a + p\underline{K}_b = p\underline{K}_W$ (the ion product of H_2O)
   As derived, this expression holds only for equilibria in H_2O.

   b. 2,4,6-Trinitroaniline is a Hammett base, so the $\underline{H}_0$ function
   can be used. The appropriate equation is

   $$p\underline{K}_a = \underline{H}_0 - \log \frac{[In]}{[InH^+]} \;.$$

   Since $[In]/[InH^+]$ is $92.9/7.1$ or 13.1,

   $$p\underline{K}_a = -8.92 - \log(13.1)$$
   $$= -10.04$$

   The $\underline{H}_0$ for 90% H_2SO_4 of -8.92 is found in any table of acidity
   functions.

   c. Although 2,4-dichloro-6-nitroaniline (A) is a Hammett base,
   2,4,6-trichlorobenzamide (B) is not. A separate acidity
   function ($\underline{H}_A$) has been developed for use with amide bases.
   The same equation as used in (b) may be applied here, with

the appropriate acidity function. For A,

$$\underline{pK}_a = \underline{H}_0 - \log \frac{1}{99}$$

$$-3.30 = \underline{H}_0 + 2$$

$$\underline{H}_0 = -5.30.$$

For an $\underline{H}_0$ of -5.30 62% H_2SO_4 must be used. For B,

$$\underline{pK}_a = \underline{H}_A - \log \frac{1}{99}$$

$$\underline{H}_A = -5.30.$$

For an $\underline{H}_A$ of -5.30, 92% H_2SO_4 must be used. For much of the acid range, $\underline{H}_A$ is more negative than $\underline{H}_0$ by 2 to 3 units. Thus aqueous H_2SO_4 is less effective by up to 1000 times in protonating an amide, which probably protonates on the -OH group, than in protonating a primary aniline of comparable $\underline{pK}_a$. The species B requires a much stronger acid solution than A for protonation to the same extent. These results point up the sensitivity of the acidity function to base structure.

References: E. B. Kelsey and H. G. Dietrich, "Fundamentals of Semimicro Qualitative Analysis," 2nd ed, Macmillan, New York, NY, 1951, pp 64-73; K. Yates, J. B. Stevens, and A. R. Katritzky, Can. J. Chem., 42, 1957 (1964); for a typical table of acidity functions, see L. P. Hammett, "Physical Organic Chemistry," 2nd ed, McGraw-Hill, New York, NY, 1970, p 271.

2. a. The acid equilibrium for $\overset{+}{In}H$ including the solvent H_2O

$$^{+}In-H + H_2O \rightleftharpoons In + H_3O^{+}$$

yields the following expression for the acidity function.

$$\underline{H}_0 = -\log\frac{a_{H_3O^{+}} \, \gamma_{In}}{\gamma_{InH^{+}}} + \log a_{H_2O}$$

Changes in acidity functions can result from the changes in any of these terms.

(i) As the acidity of a solution increases, the activity of H_2O decreases. If the above equilibrium were to take

78

explicit account of the hydration of each species, the term $\log a_{H_2O}$ would be multiplied by a coefficient $\underline{n}$ representing the difference between the hydration of the species on the right and that of those on the left of the equation. Since the acids have varying hydration numbers (as written, 3 for $Ar\overset{+}{N}H_3$, 1 for $Ar_2CH_3\overset{+}{N}H$, none for $Ar_2\overset{+}{C}CH_3$), variation is expected in the acidity functions on the basis of the term $\underline{n} \log a_{H_2O}$.

(ii)   The amount of hydration affects the activity coefficients. Thus γ_{InH^+} and to a less extent γ_{In} depend on the number of hydrogen bonds, so that the acidity functions will vary with activity coefficient changes.

Thus $Ar\overset{+}{-}NH_3$ ($\underline{H}_0$) has the greatest hydration through hydrogen bonding and produces the least negative series of $\underline{H}_X$ functions of those given. Hydration is smaller for $Ar_2CH_3\overset{+}{N}H$ ($\underline{H}'''$) and least for $Ar_2\overset{+}{C}CH_3$ ($\underline{H}'_R$), so $\underline{H}'_R$ has the most negative set of $\underline{H}_X$ values. The apparent low hydration of the carbonium ion species ($\underline{H}'_R$) has led Arnett to suggest that they undergo general dielectric solvation rather than specific hydration. Delocalization of charge renders specific solvation less necessary or possible. Almost any explanation of acidity function variation in terms of only one parameter, such as hydration, is probably simplistic. The hypothesis described in the statement of the problem, however, is useful as a first approximation.

b.   Under the present hypothesis, the extent of hydration through hydrogen bonding of ^+InH is the critical quantity in determining variations in acidity functions. Since ketones produce an acidity function similar to that of primary amines, which have three hydrogen bonds to H_2O, it appears that protonated ketones also must have three H_2O molecules attached through hydrogen bonds. At first glance, only one hydrogen bond seems possible, since only one proton is available. Additional hydration must be present in a model such as A. Again, it

A

$$
\begin{array}{c}
\text{H----OH}_2 \\
|\\
\overset{+}{\underset{\|}{O}}\!\!\!-\!\!\text{H---- O--H----OH}_2 \\
\| \\
C \\
C_6H_5 \quad C_6H_5
\end{array}
$$

should be emphasized that this model is a gross oversimplification of a complex situation. A similar explanation has been offered for why $\underline{H}_A$ (see problem 5-1) is even less negative than $\underline{H}_0$. Up to five waters of hydration have been allotted to the protonated amide on this basis.

c. For $\underline{H}_-$ functions, the charged indicator is an anion and hence is the lone pair donor in the hydrogen bond. The anions from each of the given species have essentially equivalent abilities to provide hydrogen bonding. An

$$
\text{Ar}-\overset{-}{N}\!\!\diagdown\!\!\begin{array}{c}H\\ \text{HOH}\end{array} \qquad \text{Ar}-\overset{-}{N}\!\!\diagdown\!\!\begin{array}{c}Ar\\ \text{HOH}\end{array} \qquad \text{NC}-\overset{-}{CH}\!\!\diagdown\!\!\begin{array}{c}CN\\ \text{HOH}\end{array}
$$

alternative view is that these delocalized anions interact weakly (or to about the same extent) with hydrogen-bonding solvents, whether the anion is primary, secondary, nitrogen, or carbon. The $\underline{H}_-$ scale for carbon acids was actually found to be about 0.4 units higher than the scale for nitrogen acids. This small difference may be real, or it may be an artifact of referencing. All primary and secondary amines lie on the same line; all carbon acids studied lie on a congruent line 0.4 $\underline{H}_-$ units above the nitrogen line, after correction for base strength differences. Nonetheless, it is clear from these studies that $\underline{H}_-$ functions are less sensitive than $\underline{H}_0$ functions to indicator structure.

References: C. H. Rochester, "Acidity Functions," Academic Press, London, 1970, pp 80-95; L. P. Hammett, "Physical Organic Chemistry," 2nd ed, McGraw-Hill, New York, NY,

1970, pp 281-82; D. Dolman and R. Stewart, Can. J. Chem., 45, 911 (1967); K. Bowden and A.F. Cockerill, J. Chem. Soc. B, 173 (1970).

3. a.

$$\underline{H}_{OH} = -\log \frac{a_{H^+} \, \gamma_{ROH}}{\gamma_{ROH_2^+}}$$

$$\underline{H}_R = -\log \frac{a_{H^+} \, \gamma_{ROH}}{a_{H_2O} \, \gamma_{R^+}}$$

The function $\underline{H}_R$ differs from other acidity functions in that it describes the ability of a medium to protonate an alcohol and dehydrate the oxonium ion to form a carbonium ion. The function $\underline{H}_{OH}$ measures only the ability to protonate an alcohol. Subtraction of the defining equations gives

$$\underline{H}_R - \underline{H}_{OH} = \log \frac{a_{H^+} \gamma_{ROH}}{\gamma_{ROH_2^+}} \cdot \frac{a_{H_2O} \gamma_{R^+}}{a_{H^+} \gamma_{ROH}}$$

$$= \log a_{H_2O} + \log \frac{\gamma_{ROH} \gamma_{R^+}}{\gamma_{ROH_2^+} \gamma_{ROH}} \cdot$$

No further reduction (except cancelation of γ_{ROH}) is possible without arbitrary assumptions about activity coefficients. If $\log \dfrac{\gamma_{ROH}}{\gamma_{ROH_2^+}}$ and $\log \dfrac{\gamma_{R^+}}{\gamma_{ROH}}$ are the same (the Hammett hypothesis) the activity coefficient term drops out to give

$$\underline{H}_R - \underline{H}_{OH} = \log a_{H_2O} \cdot$$

Otherwise, the simplest relationship is

$$\underline{H}_R - \underline{H}_{OH} = \log a_{H_2O} + \log \frac{\gamma_{R^+}}{\gamma_{ROH_2^+}} \cdot$$

b. The experimental observation is that $\underline{H}_R$ and $\underline{H}_{OH} + \log a_{H_2O}$ are not equivalent. Therefore the assumption in (a) that the activity coefficient ratios cancel must be incorrect. The ratio $\gamma_{R^+}/\gamma_{ROH_2^+}$ is indeed nonzero and decreases with increasing acid concentration. Thus $\underline{H}_R$ and $\underline{H}_{OH}$, which refer to the same substrate, differ formally not only in the term $\log a_{H_2O}$ but also in the activity coefficients of the cationic species. This latter difference presumably arises from the different solvation properties of oxonium $(R\overset{+}{O}H_2)$ and carbonium (R^+) ions, as follows from the discussion in problem 5-2a.

Reference: C. H. Rochester, "Acidity Functions," Academic Press, London, 1970, pp 72-80. The discussion was adapted for $\underline{H}_{OH}$ from $\underline{H}_0$.

4. a.
$$\log \frac{\gamma_{XH^+}}{\gamma_{X}\gamma_{H^+}} = (1 - \phi) \log \frac{\gamma_{InH^+}}{\gamma_{In}\gamma_{H^+}}$$

$$\log \frac{\gamma_{XH^+}[H^+]}{\gamma_{X}a_{H^+}} = (1 - \phi) \log \frac{\gamma_{InH^+}[H^+]}{\gamma_{In}a_{H^+}}$$

$$\underline{H}_X = \log \frac{\gamma_{X}a_{H^+}}{\gamma_{XH^+}} \qquad \underline{H}_0 = -\log \frac{\gamma_{In}a_{H^+}}{\gamma_{InH^+}}$$

$$\underline{H}_X + \log[H^+] = (1 - \phi)(\underline{H}_0 + \log[H^+])$$

It is normally more convenient to plot $\underline{H}_X + \log[H^+]$ vs. $\underline{H}_0 + \log[H^+]$ than to use the initial activity coefficient expressions in order to get ϕ.

b. For Hammett bases, of course, X and In are the same thing, so

$$\log \frac{\gamma_{InH^+}}{\gamma_{In}\gamma_{H^+}} = (1 - \phi) \log \frac{\gamma_{InH^+}}{\gamma_{In}\gamma_{H^+}} .$$

$$1 = 1 - \varphi$$
$$\varphi = 0$$

Thus the solvation parameter ϕ is defined for other bases by comparison with a value of zero for Hammett bases (primary aromatic amines).

c. If the base X is H_2O,

$$\log \frac{\gamma_{H_3O^+}}{\gamma_{H_2O}\gamma_{H^+}} = (1 - \phi) \log \frac{\gamma_{InH^+}}{\gamma_{In}\gamma_{H^+}} .$$

In aqueous solutions, γ_{H^+} can be taken to be $\gamma_{H_3O^+}/\gamma_{H_2O}$.

$$\log \frac{\gamma_{H_3O^+}\gamma_{H_2O}}{\gamma_{H_2O}\gamma_{H_3O^+}} = (1 - \Phi) \log \frac{\gamma_{InH^+}}{\gamma_{In}\gamma_{H^+}}$$

$$\log 1 = 0 = (1 - \Phi) \log \frac{\gamma_{In}}{\gamma_{In}\gamma_{H^+}}$$

$$\Phi = 1.0$$

This value is the extreme upper limit of Φ.

d. According to the definition given at the beginning of the problem, Φ is a measure of the relative changes of $\gamma_{XH^+}/\gamma_X\gamma_{H^+}$ vs. $\gamma_{InH^+}/\gamma_{In}\gamma_{H^+}$. A negative Φ is obtained when the former changes less with acid concentration than the latter. Thus when ^+XH has a strongly delocalized positive charge, γ_{XH^+} may not differ very much from γ_X, and the change of their ratio with acid would be small in comparison with that of $\gamma_{InH^+}/\gamma_{In}$. A large negative Φ would then result, as is the case for $\underline{H_R}$, the acidity function for formation of tertiary carbonium ions through dehydration. The same would be true for $\underline{H'_R}$. Oxonium ions (R_3O^+) require greater solvation than ammonium ions, because of the smaller size and polarizability of oxygen compared to nitrogen. Thus $\underline{H_0}$ (nitrogen acids) has an intermediate Φ, and $\underline{H_{ROR}}$ (oxygen acids) has a large positive Φ that expectedly is close to that of H_2O. Furthermore, oxygen acids have stronger hydrogen bonds to the solvent H_2O that enhance solvation. The number of hydrogen bonds is also important (see problem 5-2) (compare $\underline{H_0}$ to H'''). Sulfonium ions are more polarizable than oxonium ions, so that γ_{XH^+} is closer to γ_X and the change in their ratio is smaller. Thus $\underline{H_{RSR}}$ has a slightly negative Φ. In sum, the more concentrated the charge is on ^+XH, the greater the solvation and the larger the Φ.

References: J. F. Bunnett and F. P. Olsen, Can. J. Chem., 44, 1899 (1966); P. Bonvicini, A. Levi, V. Lucchini, G. Modena, and G. Scorrano, J. Am. Chem. Soc., 95, 5960 (1973).

5. a. From problem 5-4,

$$\underline{H}_X + \log[H^+] = (1 - \Phi)(\underline{H}_0 + \log[H^+]).$$

$$p\underline{K}_a - \log\frac{[XH^+]}{[X]} + \log[H^+] = \underline{H}_0 - \Phi\underline{H}_0 + \log[H^+] - \Phi\log[H^+]$$

$$p\underline{K}_a = \log\frac{[XH^+]}{[X]} + \underline{H}_0 - \Phi(\underline{H}_0 + \log[H^+])$$

If Φ is known, then the $p\underline{K}_a$ can be calculated from the measurement of $[^+XH]/X$ and from tables of $\underline{H}_0$.

b. Rearrangement of the final equation in (a) gives

$$\underline{H}_0 + \log\frac{[^+XH]}{[X]} = \Phi(\underline{H}_0 + \log[H^+]) + p\underline{K}_a.$$

This equation is of the form $\underline{y} = \underline{m}\,\underline{x} + \underline{b}$, so that a plot of $\underline{H}_0 + \log([^+XH]/[X])$ vs. $\underline{H}_0 + \log[H^+]$ (by varying $[H^+]$, measuring $[^+XH]/[X]$, and looking up $\underline{H}_0$) gives a slope of Φ and an intercept of the $p\underline{K}_a$. Alternative expressions with $\underline{H}_0$ in either the slope or the intercept are unacceptable, since $\underline{H}_0$ is a variable in the experiment.

c. Clearly there is no single ketone acidity function, since Φ changes enormously with modest changes in structure. The large, positive Φ, as in protonated acetone (A), means that γ_{XH^+}/γ_X is changing more rapidly than $\gamma_{InH^+}/\gamma_{In}$.

The charge on ^+XH is relatively localized and greater solvation is required. The smaller positive or larger negative Φ's are associated with increased delocalization of the positive charge, as in protonated 4-methoxybenzophenone (B). Weaker solvation (or hydrogen bonding) is required for more dispersed charges. The protonated ketones, in contrast to ammonium ($R_3\overset{+}{N}H$) or oxonium ($R_2\overset{+}{O}H$) ions, offer considerable variability in the extent of charge delocalization, so Φ has a large range. A similar situation should hold for protonated esters, amides, or alkenes, since charge delocalization can be quite variable. Each individual base would require its own acidity function. Inversion of acidities can occur as a function of acid concentration. The $\underline{pK}_a$'s in the present study refer to dilute aqueous solution. What happens at higher concentrations of H_2SO_4? In dilute solution, protonated cyclopropyl methyl ketone ($\underline{pK}_a$ = -3.27) is more acidic than protonated acetone (-2.85). Thus cyclopropyl methyl ketone is less basic, despite the superior electron-donating effect of the cyclopropane group. The greater solvation requirements of protonated acetone (larger positive Φ) overwhelm the electronic effect. As the acidity increases and the concentration of H_2O decreases, the acidities reverse. Thus the $\underline{H}_0$ value for half protonation is -7.86 for acetone but -5.96 for cyclopropyl methyl ketone, indicating higher basicity of the latter ketone in stronger acid. Knowledge of both the $\underline{pK}_a$ and Φ for a given base permits conclusions to be drawn about acidity over the full range of acidic solutions.

Reference: A. Levi, G. Modena, and G. Scorrano, <u>J. Am. Chem. Soc.</u>, <u>96</u>, 6585 (1974).

ALSO SEE PROBLEMS 10-9, 10-10, 10-12.

6
INTRODUCTION TO
KINETICS

PROBLEMS

1. The rate of nitration of benzene is pseudozeroth order when there is a large excess of HNO_3. The rate of nitration of

$$\text{rate} = \underline{k}_0$$

nitrobenzene is pseudofirst order under the same conditions.

$$\text{rate} = k_1[C_6H_5NO_2]$$

a. Derive the observed rate equations from the mechanism of electrophilic substitution.

b. Suggest a reason for the differences between the two systems.

2. a. The rate of the reaction of acetone with bromine, as measured by the uptake of bromine, is first order in acetone concentration and in hydronium ion concentration but zeroth

$$CH_3\overset{\overset{\displaystyle O}{\|}}{C}CH_3 + Br_2 \longrightarrow HBr + CH_3\overset{\overset{\displaystyle O}{\|}}{C}CH_2Br$$

order in bromine concentration. Suggest a mechanism that is consistent with such a rate law.

b. Although the reaction is first order in acid concentration, one of the products is HBr. In order to avoid an autocatalytic effect, the reaction is normally examined under buffered conditions (acetic acid/acetate ion). Under these circumstances, the rate was found to be first order in both $[H_3O^+]$ and [HOAc]. Suggest two distinct mechanisms that are consistent with these observations.

3. Which reaction in each pair has the <u>larger</u> negative entropy of activation?

a. (i)

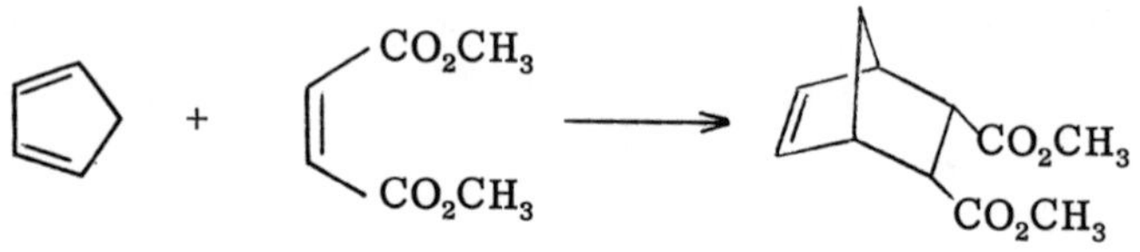

(ii)

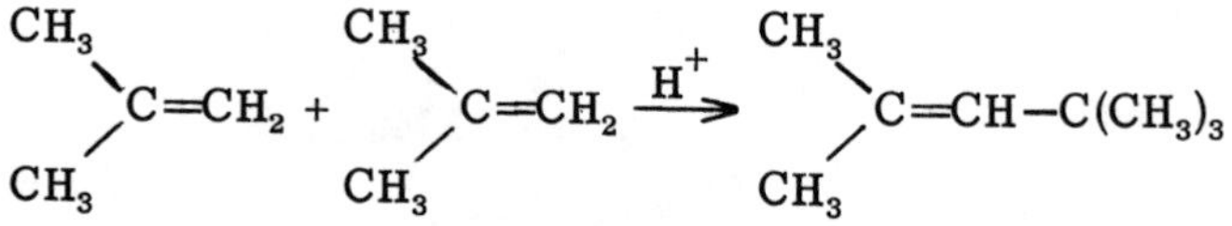

b. (i) $CH_3I + \,^-OC_2H_5 \longrightarrow CH_3OC_2H_5 + I^-$

(ii) $(CH_3)_3CCH_2I + \,^-OC_2H_5 \longrightarrow (CH_3)_3CCH_2OC_2H_5 + I^-$

c. (i) $CH_3I + NH_3 \longrightarrow CH_3\overset{+}{N}H_3 + I^-$

(ii) $CH_3I + \,^-NH_2 \longrightarrow CH_3NH_2 + I^-$

d. (i)

(ii)

e. $(C_2H_5)_3N + CH_3I \longrightarrow (C_2H_5)_3\overset{+}{N}CH_3 \quad I^-$

(i) in $C_6H_5NO_2$ as solvent

(ii) in C_6H_6 as solvent

f.   (i) $(CH_3)_2\underset{\overset{|}{CN}}{C}-N{=}N-\underset{\overset{|}{CN}}{C}(CH_3)_2 \longrightarrow N_2 + 2(CH_3)_2\overset{\cdot}{C}CN$

(ii)

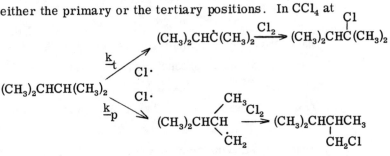

4. a. In the S_N1 hydrolysis of secondary and tertiary halides, the
intermediate carbonium ion can be partitioned between var-
ious nucleophiles. The relative values of $\underline{k}_2$ for azide ion and

$$R_3C-Cl \xrightarrow{\underline{k}_1} R_3\overset{+}{C} \underset{\underline{k}_2(Nu'')}{\overset{\underline{k}_2(Nu')}{\big\langle}} \begin{array}{l} R_3C-Nu' \\[10pt] R_3C-Nu'' \end{array}$$

water provide a measure of this partitioning process. For
$(C_6H_5)_3CCl \; \underline{k}_2(^-N_3)/\underline{k}_2(H_2O)$ is $280,000$, for $(C_6H_5)_2CHCl$ it is
170, and for $(CH_3)_3CCl$ it is 4. Account for these changes in
$\underline{k}_2(^-N_3)/\underline{k}_2(H_2O)$ with the help of an energy level diagram.

b. Free radical halogenation of 2,3-dimethylbutane can occur at
either the primary or the tertiary positions. In CCl_4 at

$$(CH_3)_2CHCH(CH_3)_2 \underset{\underline{k}_p}{\overset{\underline{k}_t}{\big\langle}} \begin{array}{l} Cl\cdot \\[6pt] (CH_3)_2CH\overset{\cdot}{C}(CH_3)_2 \xrightarrow{Cl_2} (CH_3)_2CH\overset{\overset{\textstyle Cl}{|}}{C}(CH_3)_2 \\[18pt] Cl\cdot \\ (CH_3)_2CH\overset{\overset{\textstyle CH_3}{\diagup}}{\underset{\underset{\textstyle \dot{C}H_2}{\diagdown}}{CH}} \xrightarrow{Cl_2} (CH_3)_2CHCH\underset{\overset{|}{CH_2Cl}}{CH_3} \end{array}$$

$55\,^\circ C$, $\underline{k}_t/\underline{k}_p$ is 3.5 (already corrected for the 6/1 statistical
difference), in $\underline{n}$-butyl ether it is 7.2, and in benzene it is
14.6. Account for these changes in $\underline{k}_t/\underline{k}_p$ with the help of an
energy level diagram.

5. Related subject: conformational analysis

The Curtin-Hammett principle applies to situations in which products arise from interconverting species. If the activation

$$N \longleftarrow A \rightleftharpoons B \longrightarrow M$$

energies ($\Delta G_A^\ddagger$, $\Delta G_B^\ddagger$) for the reactions of the two species are large compared to the barrier to interconversion ($A \rightleftharpoons B$), the ratio of the products from the two species (p_M/p_N) does not depend on the relative isomeric populations in the ground state (p_B/p_A) but only on the difference in the energies of the respective transition states from the two species ($G_A^\ddagger - G_B^\ddagger$).

a. Show that p_M/p_N depends only on $G_A^\ddagger$, $G_B^\ddagger$, and the temperature. Assume that B is lower in energy than A and that $G_B^\ddagger$ is lower than $G_A^\ddagger$ (these assumptions are arbitrary and do not limit the proof).

b. Consider the following elimination reaction. trans-Stilbene is formed in 99 times greater abundance than cis-stilbene (corresponding to about 3.2 kcal/mol^{-1}). Draw the two reactive conformations, assuming that elimination is anti-

$$C_6H_5CH_2\overset{\overset{\displaystyle OCOR}{|}}{C}HC_6H_5 \xrightarrow{\text{K}^+ \, {}^-O\text{-tert-}C_4H_9} \quad + \quad$$

periplanar. Which conformer is more stable and why? Does the more abundant product come from the more or the less abundant starting material?

c. trans-Stilbene is about 5.7 kcal mol^{-1} stabler than cis-stilbene. Draw a complete energy level diagram of this reaction, showing relative energies. Does the Curtin-Hammett principle apply?

d. Amino alcohols containing aryl groups can deaminate with aryl migration. The aryl group must be anti to $^+N_2$ in order to migrate. For the following diastereoisomer, the product

$$\underset{\text{Ar NH}_2}{\overset{\text{Ar CH}_3}{\text{HO}-\text{C}-\text{C}-\text{H}}} \xrightarrow{\text{HONO}} \underset{\text{Ar} \;\text{N}_2^+}{\overset{\text{Ar} \;\text{CH}_3}{\text{HO}-\text{C}-\text{C}-\text{H}}} \xrightarrow[\text{-N}_2]{\text{-H}^+} \underset{\text{H}}{\overset{\text{O Ar}}{\text{Ar}-\text{C}-\text{C}-\text{CH}_3}}$$

from migration of Ar is in great predominance. Give the

$$\longrightarrow \text{Ar'COCHCH}_3\text{Ar} + \text{ArCOCHCH}_3\text{Ar'}$$

A   B

specific conformers that lead to the two products (A and B),
and suggest whether the Curtin-Hammett principle applies.
The activation energy to deamination is about 5 kcal mol^{-1}.

6. 1-Lithio-2-methyl-1-phenylpropene (A) isomerizes to the allylic
species (B) in tetrahydrofuran above 0 °C in the presence of
β,β-dimethylstyrene (C). The reaction was found to be first
order in the hydrocarbon C and half order in the vinyllithium

C   A   B

species A. Higher concentrations of added LiBr did not depress
the rate. The activation parameters were found to be $\Delta H^{\ddagger} =$
8.2 kcal mol^{-1} and $\Delta S^{\ddagger} = -44$ eu. Account for each of these facts
in a discussion of the mechanism of this isomerization.

7. Nitrenes, like carbenes, can exist as singlet and triplet modifi-
cations. It is thought that carbethoxynitrene ($:\ddot{\text{N}}-\text{CO}_2\text{C}_2\text{H}_5$) is
generated in the singlet state by α elimination from N-(p-nitro-
benzenesulfonyloxy) urethane, but that its ground state is a
triplet.

a. Reaction of carbethoxynitrene with cis-4-methyl-2-pentene
gives an aziridine adduct whose stereochemistry is dependent

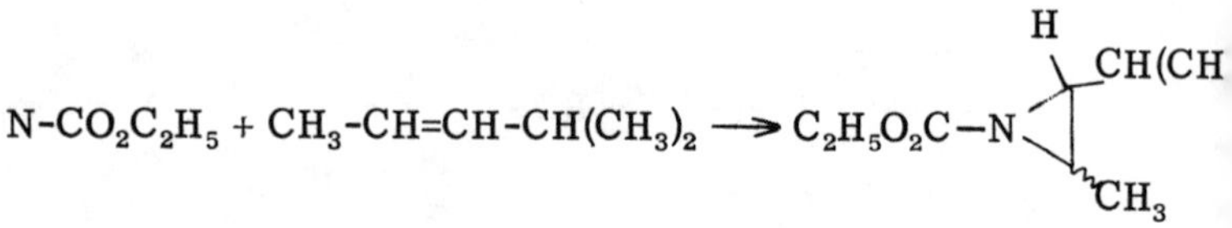

$N-CO_2C_2H_5 + CH_3-CH=CH-CH(CH_3)_2 \longrightarrow C_2H_5O_2C-N$

on the concentration of alkene in the reaction mixture. For example, the aziridine product from the cis alkene is 7.8% trans in CH_2Cl_2 containing 33% alkene, and 43% trans for 1.5% alkene. Explain.

b. Based on your explanation in (a), derive a relationship between the ratio of the respective triplet and singlet nitrene products and the concentration of alkene.

c. Reaction of carbethoxynitrene with cyclohexene gives a mixture of aziridine and cyclohexyl urethane. The proportion of urethane varies from about 14% in pure cyclohexene to 4.4%

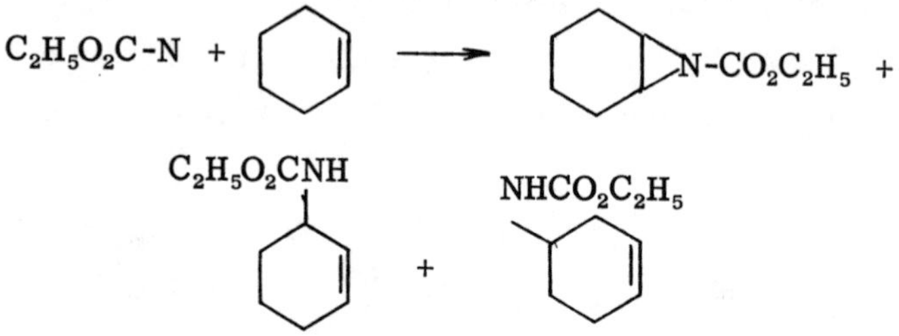

for 0.8% cyclohexene in CH_2Cl_2. From the model in (a) and (b), what can be said about the respective reactivities of singlet and triplet nitrene toward C-H insertion and C=C addition?

8. Related subject: orbital symmetry

The mechanism of free radical aromatic substitution can be

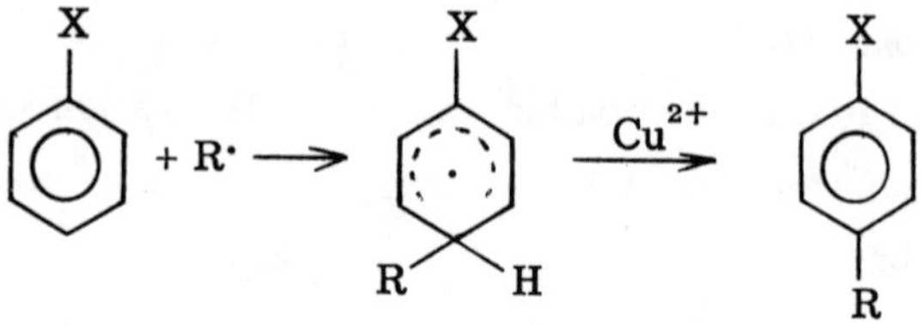

discussed usefully in terms of the concepts of frontier molecular orbital (FMO) theory.

a. Homolytic alkylation of protonated 4-methylpyridine (79 °C) is
    much faster than the corresponding reaction of anisole (k_{rel} =
    1.2 x 10^{-2}) or of neutral 4-methylpyridine (k_{rel} = 1.4 x 10^{-2}).
    Protonated 4-cyanopyridine is 16 times as fast as protonated
    4-methylpyridine. Explain these observations in terms of
    FMO theory.

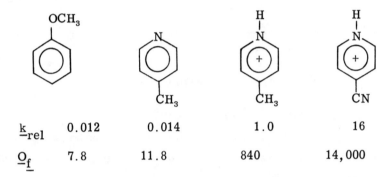

	OCH$_3$	CH$_3$ (N)	H–N$^+$ CH$_3$	H–N$^+$ CN
k_{rel}	0.012	0.014	1.0	16
O_f	7.8	11.8	840	14,000

b. The partial rate factors (O_f) for position 2 are given under
    the structures in part (a) ($\overline{O}_f$ = 1.0 for benzene). Anisole
    and neutral 4-methylpyridine, which have low reactivity
    (part (a)), also have very low selectivity. As reactivity in-
    creases with protonated 4-methyl- and 4-cyanopyridine, so
    does selectivity. Using FMO theory, suggest an explanation
    for this apparent violation of the reactivity-selectivity prin-
    ciple.

SOLUTIONS

1. a. The mechanism of nitration involves formation of the nitro-
    nium ion $^+NO_2$ and electrophilic attack of this species on the
    aromatic ring.

$$2HONO_2 \underset{k_{-1}}{\overset{k_1}{\rightleftharpoons}} H_2\overset{+}{O}NO_2 + NO_3^- \qquad \underline{K}_1 = \frac{[H_2\overset{+}{O}NO_2][NO_3^-]}{[HONO_2]^2}$$

$$H_2\overset{+}{O}NO_2 \underset{k_{-2}}{\overset{k_2}{\rightleftharpoons}} HOH + {}^+NO_2$$

$${}^+NO_2 + ArH \overset{k_3}{\longrightarrow} Ar\overset{+}{H}NO_2$$

$${}^+Ar\overset{}{H}NO_2 \overset{fast}{\longrightarrow} ArNO_2 + H^+$$

Either the second or the third step might be rate determining.
In the general case, the rate of loss of the aromatic compound
is given by

$$-\frac{d[ArH]}{dt} = \underline{k}_3[{}^+NO_2][ArH].$$

The steady state expression for the transient $[{}^+NO_2]$ is

$$\frac{d[{}^+NO_2]}{dt} = 0 = \underline{k}_2[H_2\overset{+}{O}NO_2] - \underline{k}_{-2}[H_2O][\overset{+}{N}O_2] - \underline{k}_3[\overset{+}{N}O_2][ArH].$$

Here the rate of formation of $^+NO_2$ was set equal to its rate
of destruction. Solution for $[{}^+NO_2]$ gives

$$[{}^+NO_2] = \frac{\underline{k}_2[H_2\overset{+}{O}NO_2]}{\underline{k}_{-2}[H_2O] + \underline{k}_3[ArH]} .$$

The quantity $[H_2\overset{+}{O}NO_2]$ can be replaced by the expression for
the equilibrium $\underline{K}_1$.

$$[{}^+NO_2] = \frac{\underline{K}_1\underline{k}_2[HONO_2]^2}{[NO_3^-]\,(\underline{k}_{-2}[H_2O] + \underline{k}_3[ArH])}$$

Substitution of this expression into that for the loss of aromatic compound gives

$$-\frac{d[\text{ArH}]}{dt} = \frac{K_1 \underline{k}_2 \underline{k}_3 [\text{HONO}_2]^2 [\text{ArH}]}{[\text{NO}_3^-](\underline{k}_{-2}[\text{H}_2\text{O}] + \underline{k}_3[\text{ArH}])} .$$

Since HONO_2 is in great excess, it can be assumed that $[\text{HONO}_2]$ and $[\text{NO}_3^-]$ do not change and can be subsumed into the proportionality constant.

$$-\frac{d[\text{ArH}]}{dt} = \frac{\underline{K}[\text{ArH}]}{\underline{k}_{-2}[\text{H}_2\text{O}] + \underline{k}_3[\text{ArH}]}$$

If the third step is rate determining $(\underline{k}_{-2}[\text{H}_2\text{O}] \gg \underline{k}_3[\text{ArH}])$,

$$-\frac{d[\text{ArH}]}{dt} = \frac{\underline{K}[\text{ArH}]}{\underline{k}_{-2}[\text{H}_2\text{O}]} = \underline{k}_{\text{obs}}[\text{ArH}].$$

If the second step is rate determining $(\underline{k}_{-2}[\text{H}_2\text{O}] \ll \underline{k}_3[\text{ArH}])$,

$$-\frac{d[\text{ArH}]}{dt} = \frac{\underline{K}[\text{ArH}]}{\underline{k}_3[\text{ArH}]} = \underline{k}_{\text{obs}} .$$

The observation that the rate of the reaction can vary with either the zeroth or the first power of the substrate concentration is consistent with a single mechanism with a change in the rate-determining step.

b. The change in the rate-determining step results from the relative reactivities of the substrates with the electrophile. Nitrobenzene contains the electron-withdrawing nitro group that deactivates the ring to substitution. Consequently, the third step $(^+\text{NO}_2 + \text{ArH})$ is slowed to the point of becoming rate determining, so that the rate depends on the first power of the concentration of aromatic material. For the much more reactive unsubstituted benzene ring, the third step occurs so rapidly that it is no longer rate determining. The formation of the nitronium ion in the second step then becomes rate determining, and the rate is independent of the concentration of aromatic material.

References: A. Liberles, "Introduction to Theoretical Organic

Chemistry," Macmillan, New York, NY, 1968, pp 452-57;
A. A. Frost and R. G. Pearson, "Kinetics and Mechanism,"
2nd ed, Wiley, New York, NY, 1961, pp 351-63.

2. a. The rate law is

$$-\frac{d[Br_2]}{dt} = \underline{k}_{obs}[H^+][S],$$

in which S stands for the substrate acetone. The absence of
a specific reaction component (Br_2 in this case) in the rate
law indicates that this component enters into the reaction
after the rate-determining step, as is the case with benzene
in the nitration reaction (problem 6-1a). The acid dependence
suggests an equilibrium protonation in the first step. If the
protonated substrate is converted to the enol in the rate-

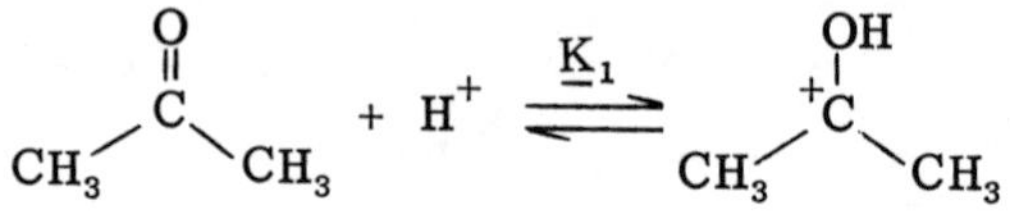

$$\underline{K}_1 = \frac{[^+SH]}{[S][H^+]}$$

determining step, the rate should be independent of $[Br_2]$.
In the general case, the step can be assumed to be reversible.

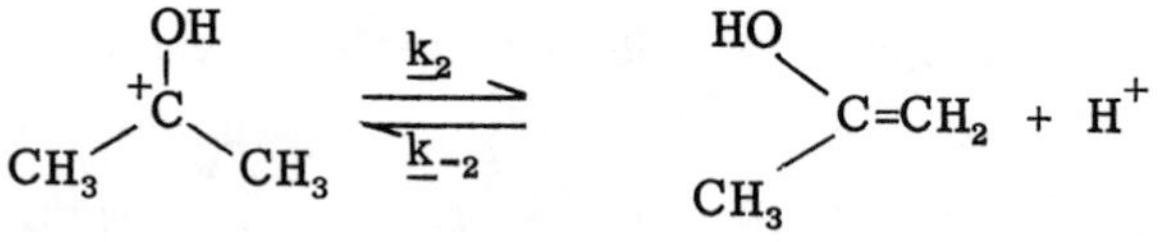

Reaction of bromine with enol then completes the reaction.

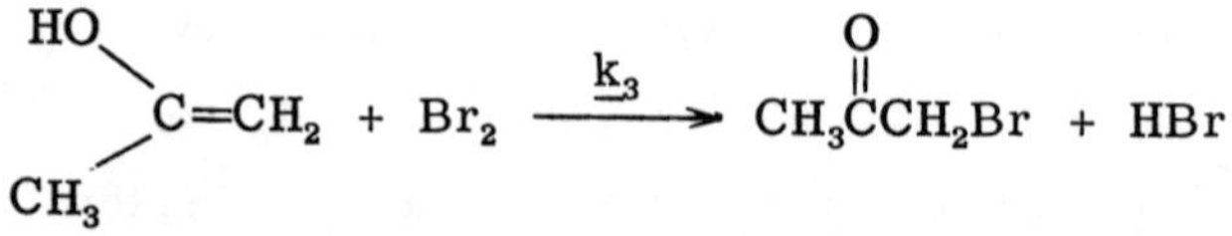

Lapworth first suggested the enol as an intermediate in the
first decade of the twentieth century. Although the kinetics
do not demand the enol, they do require rate-determining
formation of some intermediate to explain the zeroth order
dependence on $[Br_2]$. The enol pathway now seems to be on

firm grounds. According to the above three step mechanism, the loss of bromine is given by

$$-\frac{d[Br_2]}{dt} = \underline{k}_3[E][Br_2].$$

The steady state assumption for the enol (E) gives

$$\underline{k}_2[^+SH] = \underline{k}_3[E][Br_2] + \underline{k}_{-2}[E][H^+].$$

$$[E] = \frac{\underline{k}_2[^+SH]}{\underline{k}_3[Br_2] + \underline{k}_{-2}[H^+]}$$

Thus

$$-\frac{d[Br_2]}{dt} = \frac{\underline{k}_2\underline{k}_3[^+SH][Br_2]}{\underline{k}_3[Br_2] + \underline{k}_{-2}[H^+]}$$

Substitution of the acid association constant $\underline{K}_1$ gives

$$-\frac{d[Br_2]}{dt} = \frac{\underline{K}_1\underline{k}_2\underline{k}_3[S][H^+][Br_2]}{\underline{k}_3[Br_2] + \underline{k}_{-2}[H^+]} \, .$$

If the second step is rate determining $(\underline{k}_3[Br_2] \gg \underline{k}_{-2}[H^+])$, this expression simplifies to the observed rate law. If the

$$-\frac{d[Br_2]}{dt} = \underline{K}_1\underline{k}_2[S][H^+]$$

concentration of bromine is made to be extremely low $(\underline{k}_3[Br_2] \ll \underline{k}_{-2}[H^+])$, the third step can become rate determining and the rate will show a dependence on $[Br_2]$.

$$-\frac{d[Br_2]}{dt} = \underline{K}_1\underline{K}_2\underline{k}_3[S][Br_2]$$

At relatively low concentrations of bromine $(\underline{k}_3[Br_2] \sim \underline{k}_{-2}[H^+])$, the full rate expression with both terms in the denominator would be followed.

b. Catalysis by both hydronium ion and acetic acid requires the rate law

$$-\frac{d[Br_2]}{dt} = [S] \left\{ \underline{k}_{H_3O^+}[H_3O^+] + \underline{k}_{HOAc}[HOAc] \right\} \, .$$

These observations are characteristic of general acid catalysis (as opposed to specific (H_3O^+) acid catalysis). The more general rate law for general acid catalysis is

$$-\frac{d[Br_2]}{dt} = [S] \sum_i k_i [HA_i],$$

in which HA_i is a general acid and k_i is the catalytic rate constant for the ith acid. Reactions subject to general acid catalysis are also frequently subject to general base catalysis. Dawson and Spivey found that the complete rate

$$-\frac{d[Br_2]}{dt} = [S] \sum_i k_i [A_i^-]$$

law for HOAc/$^-$OAc-buffered bromination of acetone is

$$-\frac{d[Br_2]}{dt} = [S] \left\{ k_0 + k_{H_3O^+}[H_3O^+] + k_{HOAc}[HOAc] \right.$$
$$\left. + k_{-OH}[^-OH] + k_{-OAc}[^-OAc] + k_{HOAc/^-OAc}[HOAc][^-OAc] \right\},$$

in which k_0 represents the noncatalytic term (or catalysis exclusively by the solvent H_2O) and the last term is termolecular with involvement of both HOAc and $^-$OAc in the transition state. The numerical values for each of the catalytic rate constants were evaluated by Dawson and Spivey. Let us now examine possible mechanisms that would produce the general acid catalysis law. Any mechanism that introduces the full elements of HA_i (not just H^+) would give the correct kinetic expression. For example, with rate-determining transfer of a proton from HA_i to S, the first step of the acetone bromination would be

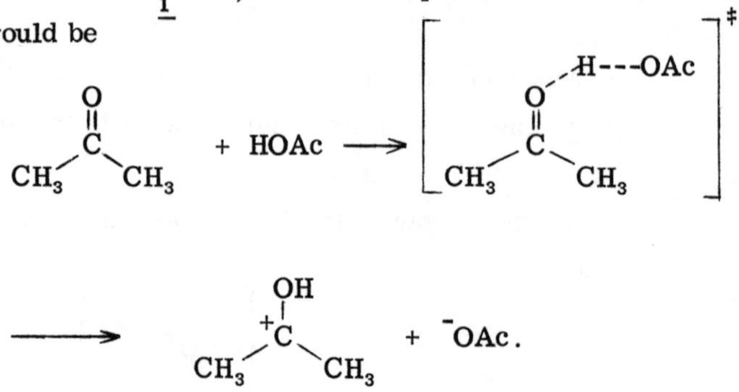

98

If this step is rate determining, the rate of reaction would be $\underline{k}_1[S][HA_i]$, as observed. A kinetically equivalent variant involves rate-determining breakdown of a complex between the substrate and the general acid. These two mechanisms will be collectively referred to as Type I General Acid Catalysis. A mechanism more closely similar to that given

Type I General Acid Catalysis

(i) $S + HOAc \longrightarrow {}^+SH + {}^-OAc$

(ii) $S + HOAc \rightleftharpoons S \cdot HOAc$

$S \cdot HOAc \rightleftharpoons {}^+SH + {}^-OAc$

in part (a) involves rapid protonation followed by rate-determining reaction with the conjugate base of the general acid to form the enol. For this mechanism, the rate (with A_i^- as

Type II General Acid Catalysis

$$S + H^+ \xrightarrow{K_1} {}^+SH$$

$$ {}^+SH + {}^-OAc \xrightarrow{k_2} E + HOAc$$

the conjugate base of the general acid) would be

$$\text{rate} = \underline{k}_2[{}^+SH][A_i^-].$$

Substitution of $\underline{K}_1$ for the first step gives

$$\text{rate} = \underline{K}_1\underline{k}_2[S][H^+][A_i^-].$$

Recognition that the acid dissociation constant for the general acid (in this case H_3O^+ or HOAc) is $\underline{K}_{ai} = [H^+][A_i^-]/[HA_i]$ and appropriate substitution gives

$$\text{rate} = \underline{K}_1\underline{k}_2\underline{K}_{ai}[S][HA_i],$$

which again is the observed general acid catalysis expression. The result is equivalent to that for rate-determining protonation. Although there is considerable mechanistic difference between rate-determining protonation by the general acid (Type I) and rapid protonation followed by rate-determining reaction with the conjugate base of the general acid (Type II), the two mechanisms are kinetically identical. A distinction can be made, however, by the use of kinetic isotope effects (see Chapters 7 and 10).

References: H. M. Dawson and E. Spivey, J. Chem. Soc., 2180 (1930) and earlier papers; A. A. Frost and R. G. Pearson, "Kinetics and Mechanism," 2nd ed, Wiley, New York, NY, 1961, pp 213-18.

3. A negative entropy of activation results from an increase in order on going from the ground state to the transition state of a reaction Increased order may result from bringing together of previously distinct elements, freezing out of conformational degrees of free-dom, or arranging of solvent molecules. Changes in solvation generally occur when charge is separated, created, or destroyed.

a. The Diels-Alder reaction (i) has the more negative $\Delta S^{\ddagger}$, since a precise arrangement of both elements in three dimensions is necessary for reaction. The acid-catalyzed polymeriza-tion of isobutylene, of which only the first step is shown, is also bimolecular but requires alignment in only two dimen-sions.

b. The degree to which conformational mobility is frozen out de-pends on the extent of crowding in the transition state. In these examples of the Williamson synthesis, the nucleophilic substitution of ethoxide ion on methyl iodide is much less crowded than that on neo-pentyl iodide. The additional three methyl groups in the neo-pentyl substrate must arrange them-selves as well as possible to minimize crowding ($\Delta S^{\ddagger} = -19.9$ eu). No such constraints are placed on the methyl sub-strate ($\Delta S^{\ddagger} = -9.5$ eu).

c. The charge types in (ii) remain the same during the progress of the nucleophilic substitution reaction. Since the charge is less dispersed in the transition state, solvation probably de-creases. On the other hand, in (i) charge separation is de-veloping in the transition state, so that solvation is increased and $\Delta S^{\ddagger}$ becomes more negative.

d. The first example of the Cope rearrangement (i) requires that the two allyl groups assume a specific rotational conformation, as shown. This increased ordering produces a negative $\Delta \underline{S}^{\ddagger}$. In (ii), much of the required ordering is built into the ground state, so that a less negative $\Delta \underline{S}^{\ddagger}$ is expected.

e. In this example of the Menschutkin reaction, charge is being separated in the transition state. Whereas relatively polar solvents such as nitrobenzene are already considerably aligned in the ground state, nonpolar but polarizable solvents such as benzene have little ground state ordering. The separation of charge in the transition state requires greater solvation for both nitrobenzene and benzene. The less polar solvent, however, experiences the greater increase in ordering ($\Delta \underline{S}^{\ddagger}$ = -41 eu for benzene, -35 eu for nitrobenzene). Although this effect is relatively small, it almost always follows this trend.

f. Unimolecular cleavage of a bond in one case (i) produces two free radicals, in the other case (ii) positive and negative charge. Whereas little solvation ordering is required in the free radical reaction ($\Delta \underline{S}^{\ddagger}$ = 10-14 eu because the number of particles increases), the separated charges require stronger solvation ($\Delta \underline{S}^{\ddagger}$ = -18.5 eu).

References: E. S. Gould, "Mechanism and Structure in Organic Chemistry," Holt, Rinehart and Winston, New York, NY, 1959, pp 181-83; K. B. Wiberg, "Physical Organic Chemistry," Wiley, New York, NY, 1964, pp 376-88.

4. a. The fundamental principle here is that the less stable intermediate is more reactive and therefore less selective. The trityl cation is the stablest of the three. Consequently, it is longest lived and has the greatest opportunity to exercise selectivity in its choice of a nucleophile for reaction. A large selectivity factor, $\underline{k}_2(^-N_3)/\underline{k}_2(H_2O)$, results. At the opposite end of the scale for this triad is the tert-butyl

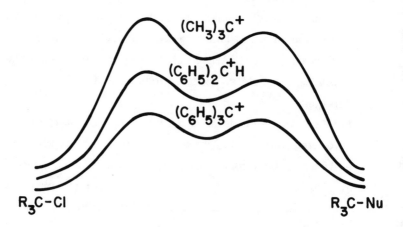

cation. Its relatively low stability means that the reaction
with a nucleophile occurs rapidly and with little discrimina-
tion, so a small selectivity factor results. Ground state
factors have been ignored. The energy level diagram shows
the relative ordering of these intermediates but omits the
branching to two transition states after the intermediate.

b.

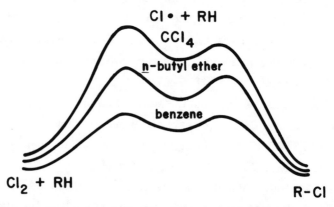

The situation is much the same here, except that a single
reaction is occurring in a sequence of solvents. Selectivity
occurs in the reaction of Cl· with the hydrocarbon. The
higher the energy of Cl· , the greater its reactivity and the
lower the selectivity. In CCl_4, Cl· is least stable and

exercises its lowest selectivity. At the other extreme, benzene must provide some stabilization of Cl·, reducing its reactivity and increasing its selectivity. The energy level diagram does not show the partitioning after the initial intermediate or the subsequent steps, which are irrelevant to the question of selectivity.

References: R. Huisgen, Angew. Chem., Int. Ed. Engl., 9, 751 (1970); R. A. Sneen, J. V. Carter, and P. S. Kay, J. Am. Chem. Soc., 88, 2594 (1966); G. A. Russell, ibid., 80, 4987 (1958); J. March, "Advanced Organic Chemistry," 2nd ed, McGraw-Hill, New York, NY, 1977, pp 630-31; for exceptions to the reactivity-selectivity principle, see B. Giese, Angew. Chem., Int. Ed. Engl., 16, 125 (1977).

5. a.

$$\frac{p_M}{p_N} = \frac{dM/dt}{dN/dt} = \frac{k_B[B]}{k_A[A]} = \frac{k_B}{k_A} K$$

In these expressions, k_B and k_A are the respective first order rate constants $\left(k_i = \frac{kT}{h} e^{-\Delta G_i^{\ddagger}/RT}\right.$, in which k is Boltzmann's constant) for the formation of M from B and of N from A, and K is the equilibrium constant $([B]/[A] = K = e^{-\Delta G^{\circ}/RT})$.

$$\frac{p_M}{p_N} = \frac{\dfrac{kT}{h} e^{-\Delta G_B^{\ddagger}/RT}}{\dfrac{kT}{h} e^{-\Delta G_A^{\ddagger}/RT}} \cdot e^{-\Delta G^{\circ}/RT}$$

$$= e^{(\Delta G_A^{\ddagger} - \Delta G^{\circ} - \Delta G_B^{\ddagger})/RT}$$

The activation energies $\Delta G_A^{\ddagger}$ and $\Delta G_B^{\ddagger}$ are positive quantities, but $\Delta G^{\circ} = G_B^{\circ} - G_A^{\circ}$ is negative as written, since B is of lower energy. Thus $\Delta G_A^{\ddagger} - \Delta G^{\circ}$ is the total energetic distance from B to the transition state of A (i.e., the sum of the A activation energy and the ground state energy differences), just as $\Delta G_B^{\ddagger}$ is the distance from B to the transition

state of B. The difference of these two positive quantities,
$(\Delta \underline{G}_A^\ddagger - \Delta \underline{G}^\circ)$ and $(\Delta \underline{G}_B^\ddagger)$, therefore is the difference in energy
of the transition states, $\underline{G}_A^\ddagger - \underline{G}_B^\ddagger$, as stated by the

$$\frac{\underline{p}_M}{\underline{p}_N} = \underline{e}^{(\underline{G}_A^\ddagger - \underline{G}_B^\ddagger)/\underline{R}\,\underline{T}}$$

Curtin-Hammett principle. The final expression contains
only $\underline{G}_A^\ddagger$, $\underline{G}_B^\ddagger$, and $\underline{T}$. The energy diagram summarizes
these relationships.

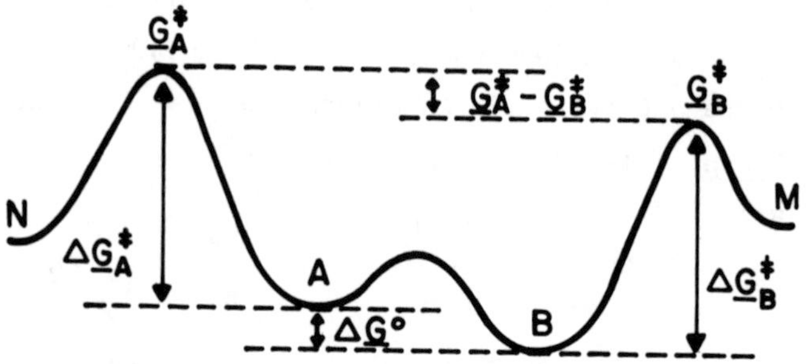

b.

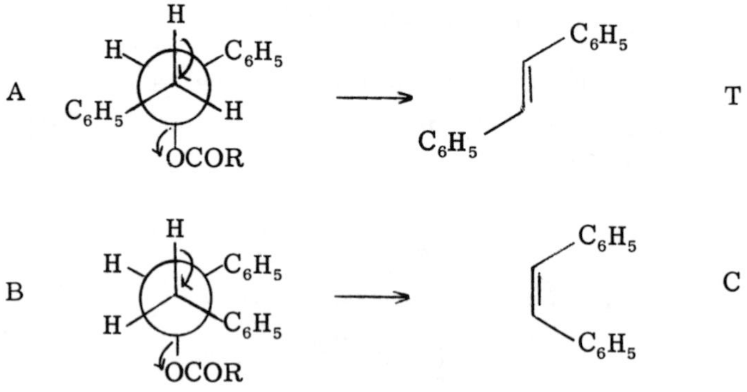

The conformer A leading to <u>trans</u>-stilbene is stabler
because the bulky phenyl groups are oriented anti to each
other. The more abundant product T does indeed come from
the more abundant conformer A.

c.

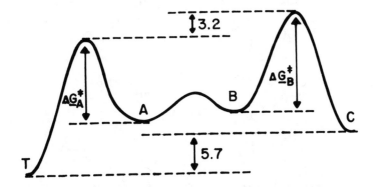

The conformational interconversion probably requires only about 5 kcal mol^{-1}, and the elimination reaction should be a much higher energy process, so one would expect that the Curtin-Hammett principle does apply. If so, $\underline{G}_A^{\ddagger} - \underline{G}_B^{\ddagger}$ is therefore 3.2 kcal mol^{-1}, from the difference in population of the products. The phenyl groups are gauche in B and synperiplanar in the product C. Interactions between the phenyl groups increase in the transition state and still more in the product. In the trans series, the antiperiplanar relationship is maintained throughout the reaction. There-fore, the cis activation energy (with increased phenyl inter-action) should be larger than the trans activation energy (with constant phenyl interaction), i.e., $\Delta\underline{G}_B^{\ddagger} > \Delta\underline{G}_A^{\ddagger}$, as the diagram shows. Because of the changing phenyl-phenyl interaction, the ground state difference ($\Delta\underline{G}^{\circ} = \underline{G}_B^{\circ} - \underline{G}_A^{\circ}$) should be smallest, the transition state difference ($\underline{G}_B^{\ddagger} - \underline{G}_A^{\ddagger}$ = 3.2 kcal mol^{-1}) should be intermediate, and the product difference ($\underline{G}_C^{\circ} - \underline{G}_T^{\circ}$ = 5.7 kcal mol^{-1}) largest.

d. Both the ground state and the transition state leading to A are lower in energy than those leading to B. In A, the Ar' is pushed toward the H in the transition state; in B the Ar is pushed toward the CH$_3$. In the ground state leading to A, the small H is between the two largest groups (Ar and Ar');

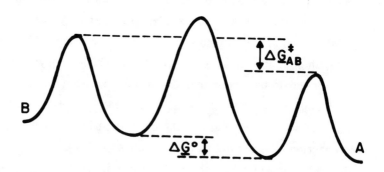

in that leading to B, the CH_3 is between them. It appears that ground state energies are determining product popula- tions, but the conclusion is not certain (see part (c)). In con- trast to the elimination reaction in (c), the loss of N_2 in the deamination is very rapid and requires little energy. The barriers to conformer interconversion and deamination proba- bly are comparable, so the Curtin-Hammett principle does not apply.

Reference: E. L. Eliel, "Stereochemistry of Carbon Compounds," McGraw-Hill, New York, NY, 1962, pp 144-45, 149-56, 237-39.

6. The first order kinetic dependence of the isomerization on the concentration of the alkene C eliminates an intramolecular

1,3-sigmatropic mechanism. The half order dependence on the concentration of the vinyllithium A requires some sort of pre-equilibrium. Dissociation of the vinyllithium into free carbanion ($RLi \rightleftharpoons R^- + Li^+$) seems unlikely because of the lack of rate retardation by added LiBr. The known aggregation of organo-lithium reagents in THF suggests that deaggregation may be the preequilibrium process. For example, breakdown of an organo-lithium dimer to the monomer and reaction of the monomer with the hydrocarbon would give overall 3/2 order kinetics.

$$R_2Li_2 \xrightleftharpoons{K_1} 2RLi$$

$$RH + RLi \xrightarrow{k_2} R'Li$$

$$K_1 = \frac{[RLi]^2}{[R_2Li_2]}$$

$$-\frac{d[RH]}{dt} = k_2[RH][RLi]$$

$$= K_1^{\frac{1}{2}}k_2[RH][R_2Li_2]^{\frac{1}{2}}$$

The actual order of the aggregate is not specified by these experiments. Any R_xLi_x unit, such as a tetramer, could dissociate to a pair of $R_{x/2}Li_{x/2}$ units, which react directly with the hydrocarbon C. Although the reaction is first order in C, only a catalytic amount of C need be present, since the reaction regenerates a molecule of C. The large negative entropy of activation suggests that the transition state is more polar and requires a larger extent of solvent immobilization than the ground state. A twentyfold decrease in the rate on change of the solvent from THF to benzene is consistent with this suggestion. Charge build-up as the carbon-lithium bond is partially dissociated would provide the increased transition state polarity.

Reference: R. Knorr and E. Lattke, Tetrahedron Lett., 4655, 4659 (1977).

7. a. The product from the singlet nitrene should be formed stereospecifically, so that the cis alkene would give only

cis aziridine. Intersystem crossing to the triplet mixture, however, would provide a nonstereospecific product. At high alkene concentration, the singlet rapidly encounters the cis alkene and reacts to form cis aziridine. As the alkene concentration is lowered, the time span before the nitrene encounters alkene increases, so that intersystem crossing produces more triplet nitrene. Because this triplet nitrene reacts nonstereospecifically, larger amounts of trans aziridine are formed at lower alkene concentrations.

b. The mechanistic scheme described in (a) can be represented by the following chart.

$$\text{precursor} \xrightarrow{\underline{k}_1} \text{singlet nitrene (}^1\text{N)} \xrightarrow{\underline{k}_2} \text{triplet nitrene (}^3\text{N)}$$

$$\underline{k}_3 \downarrow \text{alkene(A)} \qquad\qquad \underline{k}_4 \downarrow \text{alkene (A)}$$

$$\begin{array}{cc} \text{stereospecific} & \text{nonstereospecific} \\ \text{product (}^1\text{P)} & \text{product (}^3\text{P)} \end{array}$$

The rates of product formation are given by

$$\frac{d[^3P]}{d\underline{t}} = \underline{k}_4[^3N][A]$$

and

$$\frac{d[^1P]}{d\underline{t}} = \underline{k}_3[^1N][A].$$

Hence the ratio of products from the respective nitrenes is

$$\frac{[^3P]}{[^1P]} = \frac{d[^3P]/d\underline{t}}{d[^1P]/d\underline{t}} = \frac{\underline{k}_4[^3N]}{\underline{k}_3[^1N]}.$$

The steady-state production of triplet nitrene requires that

$$\frac{d[^3N]}{d\underline{t}} = \underline{k}_2[^1N] - \underline{k}_4[^3N][A] = 0.$$

Consequently, $[^3N]/[^1N] = \underline{k}_2/\underline{k}_4[A]$
and

$$\frac{[^3P]}{[^1P]} = \frac{k_2}{k_3[A]} \ .$$

In fact, a plot of the ratio of triplet products to singlet products, corrected for side reactions, vs. the reciprocal of alkene concentration was observed to be linear. The ratio $[^3P]/[^1P]$ can be derived from the observed stereospecifities, as described in the references.

c. At high alkene concentrations, the products should derive predominantly from singlet nitrene. Thus singlet nitrene must be able to undergo both C=C addition (mainly) and C-H insertion. As the concentration of the triplet increases with decreased alkene concentration, the proportion of urethane goes down. The triplet nitrene must have a higher preference for C=C addition over C-H insertion than does the singlet. In carbene chemistry, it has been reported that triplet CH_2 does not insert into the C-H bond of cyclohexene at all.

It should be emphasized that these experiments do not actually prove the existence of singlet and triplet carbethoxynitrene. Phenomenologically, the experiments suggest the existence of a "stereospecifically adding carbethoxynitrene" and a "nonstereospecifically adding carbethoxynitrene."

Reference: W. Lwowski and F. P. Woerner, J. Am. Chem. Soc., 87, 5491 (1965); J. S. McConaghy, Jr., and W. Lwowski, ibid., 89, 2357 (1967).

8. a. In the FMO model for many free radical reactions, the dominant interaction is between the singly occupied MO (SOMO) of the (in this case) nucleophilic alkyl radical and the lowest unoccupied MO (LUMO) of the closed shell substrate. Such explanations are similar to ones based on polar effects, because the HOMO and SOMO energies are related to the ionization potentials of the alkyl radical and the substrate.

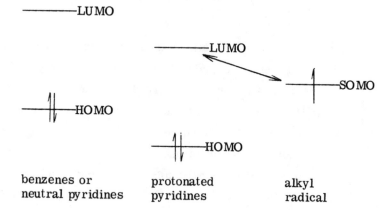

benzenes or   protonated   alkyl
neutral pyridines   pyridines   radical

Benzene, substituted benzenes, and neutral pyridines have
relatively high energy LUMO's and HOMO's. Thus the
important FMO interaction between the LUMO and the SOMO
is weak. Protonation of pyridines lowers both LUMO's and
HOMO's, permitting a stronger SOMO-LUMO interaction and
causing a faster reaction. Electron-withdrawing groups such
as cyano also lower the HOMO and the LUMO, so that the
SOMO-LUMO interaction is further strengthened and the
reaction is further accelerated.

b. Whereas reactivity is determined by the difference in energy
between the SOMO and the LUMO, regioselectivity depends
on the frontier orbital atomic coefficients. Protonated
pyridines have much higher LUMO coefficients at the ortho
and para positions than at the meta position. The differences
are not so great for benzenes and neutral pyridines. The
authors point out that as long as FMO considerations are
dominant ("when polar effects [on free radical reactions] play
a preeminent role in determining reaction rates"), the
reactivity-selectivity principle can be expected to be inappli-
cable.

References: A. Citterio, F. Minisci, O. Porta, and G. Sesano,
J. Am. Chem. Soc., 99, 7960 (1977); I. Fleming, "Frontier

Orbitals and Organic Chemical Reactions," Wiley-Interscience, London, 1976, pp 67-68, 191-94. The author is grateful to F. D. Lewis for suggesting this problem.

ALSO SEE PROBLEMS 7-1, 7-5, 7-9, 9-2, 9-3, 9-7, 9-10, 9-11, 10-2, 10-3, 10-4, 10-5, 10-6, 10-7, 10-8, 10-11, 11-6, 11-8.

7
KINETIC ISOTOPE EFFECTS

PROBLEMS

1. Related subject: introduction to kinetics

Iodination of azulene to form 1-iodoazulene follows the rate law

$$\frac{d[AzI]}{dt} = \frac{[AzH][I_2]}{[I^-]} \sum_i k_i [B_i],$$

in which AzI is 1-iodoazulene, AzH is azulene, and B_i is a general base. If deuterium is placed in the 1 position of azulene,

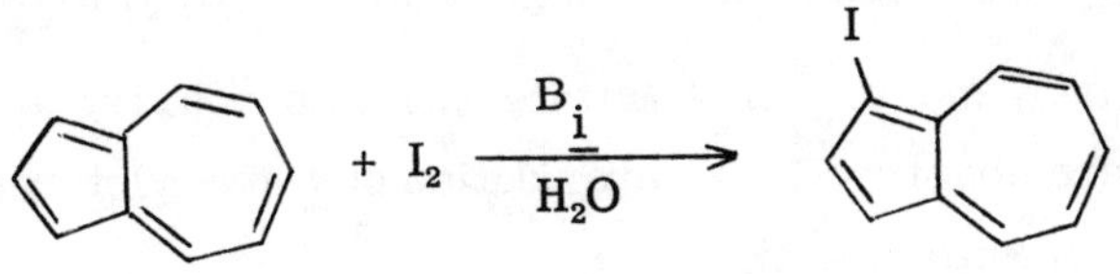

k_H/k_D is observed to be in the range 2-6.5 depending on the identity of B_i.

  a. Write out the mechanism of iodination (the standard electrophilic substitution mechanism) and specify the rate-determining step. Justify your conclusions in terms of k_H/k_D.

  b. Derive the observed rate law.

2. Free radical halogenation of toluene gives the listed kinetic isotope effects, as determined by the intramolecular preference for

$$k_H/k_D$$

$$C_6H_5CH_2D + Cl\cdot \longrightarrow C_6H_5\overset{\cdot}{C}H_2 + DCl \qquad 1.3$$

$$C_6H_5CH_2D + Br\cdot \longrightarrow C_6H_5\overset{\cdot}{C}H_2 + DBr \qquad 4.6$$

H or D, statistically corrected. Electrophilic halogenation of various aromatic species gives the listed kinetic isotope effects,

$$k_H/k_D$$

$$C_6X_6 + Cl^+ \longrightarrow C_6X_5Cl + X^+ \qquad 1.0$$

$$C_6X_6 + Br^+ \longrightarrow C_6X_5Br + X^+ \qquad 1.0\text{-}2.6$$

$$C_6X_6 + I^+ \longrightarrow C_6X_5I + X^+ \qquad 1.4\text{-}5.4$$

as determined by the separate measurement of the rates for the protonated (X = H) and deuterated (X = D) substrates. The iodine reaction is subject to general base catalysis. In both series, k_H/k_D increases with the atomic number of the halogen. Discuss the possible causes of this increase for each reaction.

3. Peracids such as m-chloroperbenzoic acid epoxidize alkenes in a synthetically useful reaction. The mechanism of the reaction

$$CH_2=CH_2 + ArCO_3H \longrightarrow \overset{\displaystyle O}{CH_2-\!\!\!-\!\!\!-CH_2} + ArCO_2H$$

is not entirely clear. The following kinetic secondary isotope effects were observed in the epoxidation of p-phenylstyrene by m-chloroperbenzoic acid.

	k_H/k_D
Ar-CD=CH$_2$	0.99
Ar-CH=CD$_2$	0.82
Ar-CD=CD$_2$	0.82

a. Suggest a mechanism that is consistent with these data. What mechanisms are excluded?

b. If $\underline{m}$-Cl-C$_6$H$_4$-CO$_3$D is the reagent for the reaction, a k_H/k_D for the same substrate (fully protonated) of 1.17 is observed. Discuss this result in light of your conclusions in part (a).

c. The above substrate ($\underline{p}$-phenylstyrene) is an unsymmetrically substituted alkene. The relative rates of epoxidation of stilbene (C$_6$H$_5$-CH=CH-C$_6$H$_5$), 4-methoxystilbene, and 4,4'-dimethoxystilbene are 1.0, 4.6, and 19.3. The substrate NO$_2$-C$_6$H$_4$-CD=CH$_2$ reacts 13 times more slowly than $\underline{p}$-phenylstyrene and shows a k_H/k_D of 0.98. What do these results say about the influence of substrate symmetry on the mechanism of the reaction?

4. Related subject: conformational analysis

A k_H/k_D close to 2 was observed in the solvolysis of (CH$_3$)$_2$-CDCHCH$_3$OTs. Since this value is too large for a secondary isotope effect, it was thought that the hydrogen is partially shifted in the transition state. A similar process might occur in the

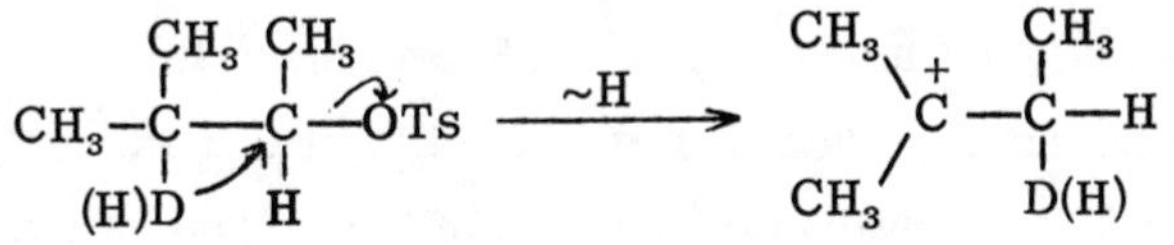

solvolysis of cyclohexyl tosylates. The following rates and isotope effects were measured for the indicated substituted cyclohexyl tosylates in 70% ethanol at 40 °C.

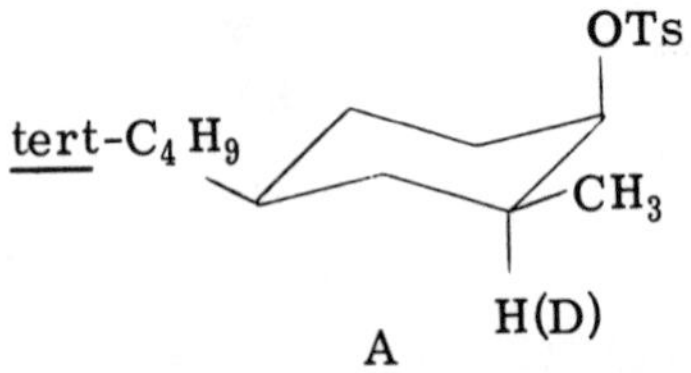

	A	B	C	D
k_H/k_D	2.08	1.96	1.19	1.15
k_H	2.775×10^{-4}	2.636×10^{-4}	1.690×10^{-6}	1.804×10^{-5}

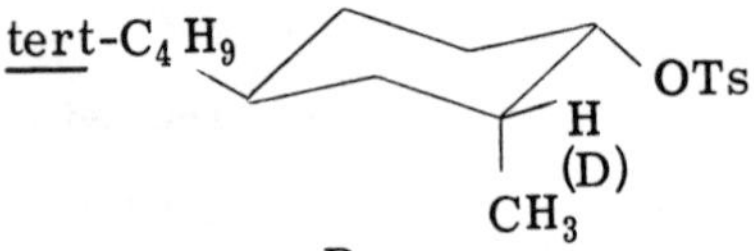

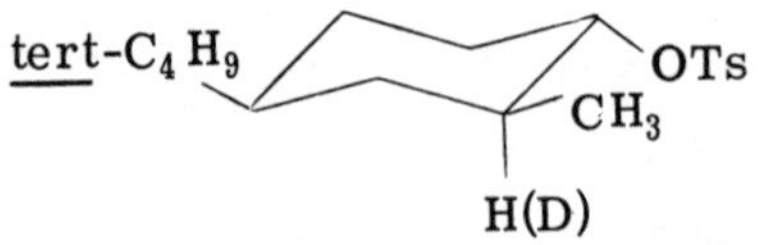

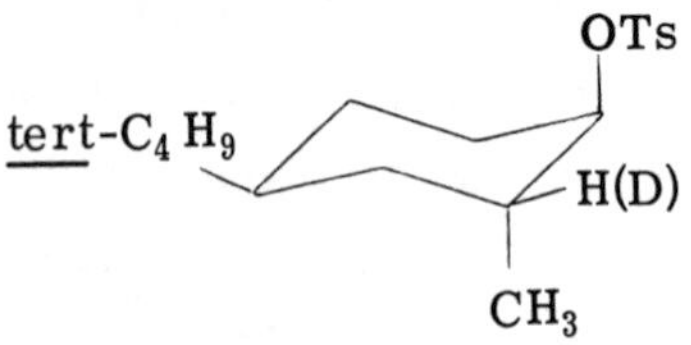

a. Why do C and D fall into one grouping with respect to k_H/k_D, and A and B into another?

b. What is probably the best geometry for participation of the hydrogen in the departure of the tosylate?

c. What special conformational effect must be invoked to explain the large k_H/k_D for B?

5. Related subjects: conformational analysis, introduction to kinetics

Base-catalyzed isomerization of 3-cyclohexenone to 2-cyclo-hexenone has a solvent isotope effect k_H/k_D (HPO_4^{2-}/H_2O to DPO_4^{2-}/D_2O) of 7.7. Deuterium exchange of the α protons (on the side of

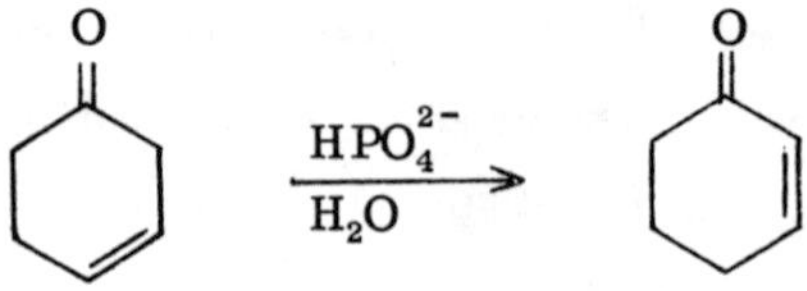

the double bond) is 575 times faster than the isomerization. For 3-cyclopentenone, k_H/k_D is 0.9 (same conditions), and α hydrogen exchange occurs at about the same rate as isomerization.

116

a. Draw out the mechanism for both systems and label the rate-determining step in each case. Show your reasoning. Derive the rate laws for these reactions.

b. 1,3-Cyclopentadiene is planar. The dihedral angle between the planes of the two double bonds in 1,3-cyclohexadiene is about $18°$. From this structural information, suggest a reason for the above differences between the cyclohexenone and cyclopentenone systems.

6. Pyridine catalysis in the synthesis of certain organic dyes is of importance to the industrial production of dyes. Consider the reaction of sulfonic acids A and B with diazonium salts $(Ar-N \overset{+}{=} N:)$. Coupling to form the azo dye ($Ar-N=N-Ar'$) occurs

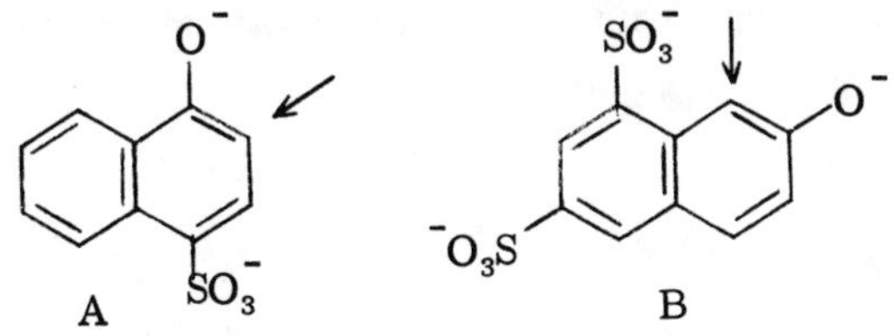

A   B

preferentially at the positions indicated by an arrow. For A, no pyridine catalysis is observed, and k_H/k_D (D at the 2 position) is 0.97. Substrate B reacts (uncatalyzed) 0.018 times as fast as A, and k_H/k_D is 6.55 (D at the 1 position). Furthermore, the reaction of B is catalyzed by pyridine, e.g., a hundredfold rate increase with 0.25 M pyridine.

a. Explain each of these differences (rate, isotope effect, pyridine catalysis) in terms of the generalized mechanism of electrophilic substitution.

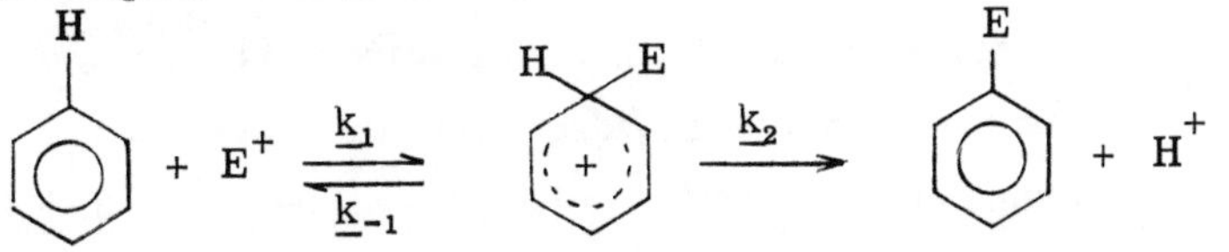

b. For substrate B, k_H/k_D is 6.55 in the absence of pyridine, 6.01 with 0.232 M pyridine, and 3.62 with 0.905 M pyridine. Explain these results in terms of your mechanism in (a). Neglect secondary isotope effects.

7. a. α-Deuterium isotope effects have been widely used to study the mechanisms of solvolysis reactions (R_2CHX vs. R_2CDX). The maximum value of about $k_H/k_D = 1.22$ for the trifluoroacetolysis of 2-propyl brosylate has been considered to be diagnostic of the limiting mechanism (k_C, free carbonium ion). Increased backside bonding to the α carbon in the transition state produces reduced isotope effects. The value of 1.138 for the ethanolysis of 3-pentyl brosylate has been interpreted in terms of the rate-determining formation of an ion pair. The overall k_H/k_D can be dissected into isotope ratios for each product, k_{iH}/k_{iD}, by adjustment for changes in product mole fractions on isotopic substitution. The following data refer to the solvolysis of 2-octyl brosylate in 65% aqueous ethanol.

Product	Yield (H species)	k_{iH}/k_{iD}
Overall reaction	100%	1.121
2-Octanol	50.50	1.108
2-Octyl ethyl ether	29.04	1.103
trans-2-Octene	11.41	1.184
cis-2-Octene	7.02	1.178
1-Octene	1.60	1.193

All previous interpretations have used only the overall k_H/k_D. Comment on how the use of dissected isotope effects will alter such interpretations. Do all the products follow from a single rate-determining step?

   b. Although rate ratios normally decrease with increased temperature, the β k_H/k_D for solvolysis reactions have been observed to increase, decrease, or remain constant. One anomalous case is the solvolysis of 2-octyl-3,3-d_2 brosylate in 65% aqueous ethanol, for which k_H/k_D remains constant (1.332 at 54 °C, 1.333 at 73 °C). Consider the dissected isotope effects. What is the mechanistic meaning of the constancy of k_H/k_D (overall)? How can k_H/k_D (overall)

Product	$\underline{k}_{iH}/\underline{k}_{iD}$ (54 °C)	$\underline{k}_{iH}/\underline{k}_{iD}$ (73 °C)
Overall reaction	1.332	1.333
2-Octanol	1.212	1.203
2-Octyl ethyl ether	1.214	1.195
trans-2-Octene	2.266	2.156
cis-2-Octene	2.289	2.179
1-Octene	1.186	1.165

remain constant when all the $\underline{k}_{iH}/\underline{k}_{iD}$ decrease with temperature?

8. Related subject: orbital symmetry

Singlet oxygen reacts with alkenes to give both cycloaddition (oxetanes from a [2 + 2] reaction) and the ene reaction (allylic hydroperoxides from a [2 + 2 + 2] reaction). Consider the following two examples. Compound B cannot undergo an ene reaction because of the absence of an allylic proton.

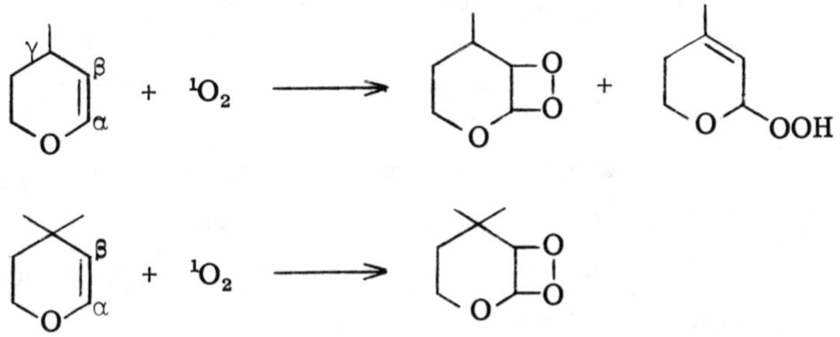

a. Conversion of an sp² carbon to an sp³ carbon normally pro-duces an inverse isotope effect, i.e., $\underline{k}_H/\underline{k}_D < 1.0$ (see problem 7-3). The following tritium kinetic isotope effects were observed in benzene for the above reactions. Com-pound A gives about 25% dioxetane in this solvent. Tritium isotope effects can be converted to deuterium isotope effects by the relationship $\underline{k}_H/\underline{k}_T = (\underline{k}_H/\underline{k}_D)^{1.442}$. What conclusions can be drawn from these data about the structure(s) of the transition state(s) to cycloaddition and ene reaction?

Compound	$k_H/k_T(\alpha)$	$k_H/k_T(\beta)$	$k_H/k_T(\gamma)$
A	0.980	0.908	1.335
B	0.994	0.897	---

b. The tritium isotope for the γ proton in A can be dissected into separate isotope effects for cycloaddition (k_H/k_T = 1.220) and ene reaction (1.376) by correction for exact product ratios. What further mechanistic conclusions can be drawn?

9. Related subject: introduction to kinetics

Azulene-1-carboxylic acid decarboxylates under acidic ($HClO_4$) conditions. Below $[H_3O^+] = 0.30$ M, the rate of decarboxylation is first order in $[H_3O^+]$. From 0.3 to 6 M, the rate of

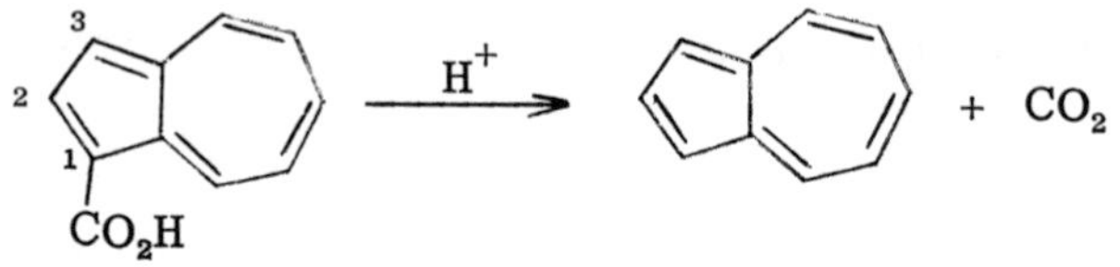

decarboxylation is independent of $[H_3O^+]$. At even higher acidities, decarboxylation fails to occur.

a. Write out a mechanism and derive the rate law for the decarboxylation.

b. Rate-determining rupture of a carbon–carbon bond is accompanied by a maximal heavy atom isotope effect of about 4% when one ^{12}C is replaced by ^{13}C, i.e., $k_{12}/k_{13} \sim 1.04$. At $[H_3O^+] = 0.01$ M, the observed k_{12}/k_{13} is about 1.004 (0.4%). At $[H_3O^+] = 0.3$ M, the value is about 1.039 (3.9%). Rationalize these results in terms of the mechanism from part (a).

SOLUTIONS

1. a. The standard mechanism of electrophilic addition involves
        reaction of the substrate with the electrophile to form a
        charge-delocalized intermediate that loses a proton to give
        the product. The rather large k_H/k_D (2-6.5) indicates that

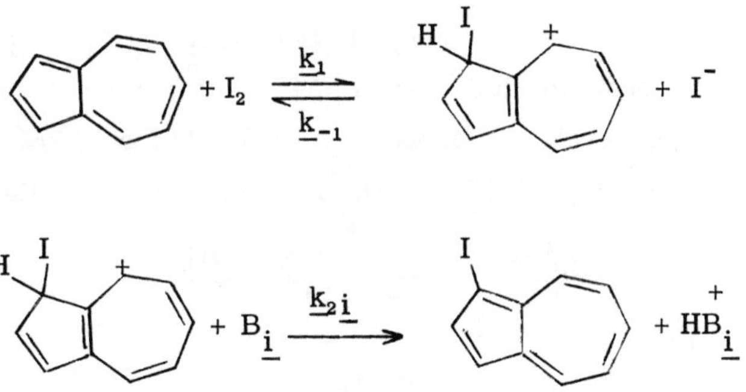

the isotope effect is primary. Cleavage of the C-H bond
occurs in the second step (reaction of the intermediate with
the general base), which must then be rate determining. See
problem 6-2 for a discussion of what is meant by a general
base. Most of the bases (H_2O, ^-OAc, pyridine among others)
have a k_H/k_D in the range 2-3.2. This relatively low primary isotope effect may be the result of a very nonlinear or
unsymmetrical transition state, or k_{2i} may not be entirely
rate determining. The abnormally large k_H/k_D for 2,4,6-
trimethylpyridine (6.5) can result either from proton tunnel-
ing in the transition state or from a steric effect on the zero
point energy of the stretched transition state (see the reference for a discussion of steric amplification of isotope effects).

    b. The overall loss of substrate is given by

$$\frac{d[AzH]}{dt} = \sum_i k_{2i}[\overset{+}{Az}HI][B_i].$$

This expression is summed over all the general bases B_i , each with its own catalytic rate constant k_{2i}. The steady state assumption can be made for the charged intermediate, ^+AzHI. The resulting expression simplifies to the

$$k_1[AzH][I_2] = k_{-1}[A\overset{+}{z}HI][I^-] + k_{2i}[\overset{+}{Az}HI][B_i]$$

$$\frac{d[AzH]}{dt} = \sum_i k_{2i}[B_i]\frac{k_1[AzH][I_2]}{k_{-1}[I^-]+k_{2i}[B_i]}$$

observed rate law, provided that $k_{-1}[I^-] \gg k_{2i}[B_i]$. This inequality holds when $[I^-] > 0.04$ M for solvent H_2O as the general base or when $[I^-] = 0.1$ M for other general bases at $[B_i] = 0.15$ M. The reduced expression then is

$$\frac{d[AzH]}{dt} = \sum_i k_{2i}[B_i]\frac{k_1[AzH][I_2]}{k_{-1}[I^-]}$$

or

$$= \frac{[AzH][I_2]}{[I^-]}\sum_i k_i[B_i],$$

in which the catalytic rate constant k_i corresponds to $k_1 k_{2i}/k_{-1}$.

Reference: E. Grovenstein, Jr., and F.C. Schmalstieg, J. Am. Chem. Soc., 89, 5084 (1967).

2. In the free radical halogenation, the isotope effect is measured intramolecularly, so there are no zero point energy differences in the ground state. The isotope effects therefore arise solely from zero point energy differences in the transition state. Decreased or small values of k_H/k_D can have at least three causes.

(a) The C-H bond is not broken in the rate-determining step.

(b) The transition state is nonlinear. This situation can occur,

Linear   C—H---X

Nonlinear

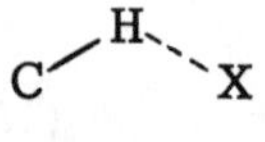

for example, in intramolecular hydrogen migrations. The stretching vibration is supplanted by bending vibrations as the important determinant of zero point vibrational energy.

(c) The transition state is unsymmetrical, that is, the force

Symmetrical $\quad\quad$ C$\overset{k_a}{-}$$-$$-$$-H\overset{k_b}{-}$$-$$-$$-$X

Unsymmetrical $\quad\quad$ C$-$$-H\cdots\cdots$X

constant for C$-$H stretching (k_a) in the transition state is
not equal to that (k_b) for H$-$X stretching.

In the free radical halogenation, there is only one step, so C$-$X
bond breaking must be rate determining (a). There is no reason
to believe that a simple hydrogen abstraction (C$-$H(D) + Cl$\cdot$) is
nonlinear (b), so the difference between chlorination and bromination must be due to the symmetry of the transition state (c). For
chlorination, the very small isotope effect (1.3) indicates that
the H$-$X bond is either very fully $(k_b > k_a)$ or very little $(k_a > k_b)$
developed in the transition state. A kinetic isotope effect this
small may in fact be secondary rather than primary. For
bromination, the respectable isotope effect (4.6) indicates that
the H$-$X bond making and the C$-$H bond breaking must be about
equal $(k_a \sim k_b)$. In the electrophilic halogenation, again there is
no reason to believe that the transition state is nonlinear (b) for
the cases with very small kinetic isotope effects. The major
difference from the free radical mechanism is that the reaction
requires two steps, either of which might be rate determining. A

primary isotope effect occurs only if the second step is rate determining. This is the usual explanation (a) given for the variation of k_H/k_D with halogen for the electrophilic reaction. The rate law for this mechanism would be

$$\frac{d[ArX]}{dt} = \frac{k_1 k_2 [ArH][X^+][B]}{k_{-1} + k_2 [B]} .$$

For chlorine, the term $k_2[B]$ must be greater than k_{-1}, so the first step is rate determining and no isotope effect is observed. For iodine, k_{-1} must be greater than $k_2[B]$, so the second step is rate determining and a significant isotope effect and base catalysis (see problem 7-1) are observed. Bromine shows intermediate behavior ($k_2[B] \sim k_{-1}$). The different abilities of Cl^+, Br^+, and I^+ to serve as leaving groups (k_{-1}) with respect to H^+ (k_2) determine which step is rate determining. The small k_H/k_D again may have an important contribution from secondary isotope effects. The isotope effect data alone are not unambiguous, since the decreased values can also result from symmetry changes in a common transition state (c).

References: K. B. Wiberg and L. H. Slaugh, J. Am. Chem. Soc., 80, 3033 (1958); E. Berliner, Progr. Phys. Org. Chem., 2, 253 (1964); R. D. Gilliom, "Introduction to Physical Organic Chemistry," Addison-Wesley, Reading, MA, 1970, pp 129-30.

3. a. A change in hybridization from sp^2 to sp^3 is normally accompanied by an inverse secondary deuterium isotope effect ($k_H/k_D < 1$). In the epoxidation reaction, both sp^2 carbons of the alkene assume a higher p character (not fully sp^3) in the oxirane product. The k_H/k_D of 0.82 for the β,β-d_2 derivative therefore is consistent with a significant hybridization change at the β carbon. The k_H/k_D of 0.99 for the α-d derivative and the unchanged k_H/k_D of 0.82 for the α,β,β-d_3 derivative indicate that little hybridization change has occurred at the α carbon. Therefore, the transition state must have a well-developed C_β-O bond but little or

no C_α-O bond, as in A. The data appear to exclude several

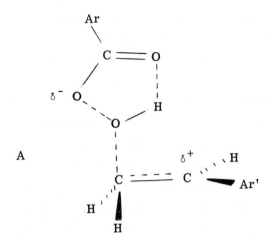

A

previously suggested mechanisms, such as the symmetrical process B and the 1,3-dipolar addition C. Undoubtedly, these mechanisms originally were considered likely because of the well-known stereospecificity of the epoxidation

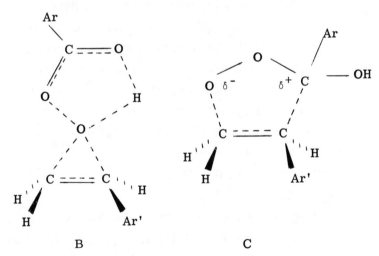

B   C

reaction. For transition state A to deliver a stereospecific product, rotation around the partial C ===C double bond must be slower than formation of the second bond to close the oxirane ring. This situation would only require the retention of a small amount of π bond character between C_α and C_β.

b. The hydroxyl (OH/OD) isotope effect is negligible. Thus in transition state A, the O—H bond is not in the process of being broken. This result is perfectly reasonable for process A and further explains why the reaction is not subject to acid catalysis. The O—H bond would amost certainly be partially broken in the transition state B. Despite the obvious nonlinearity on the O—H---O system, a k_H/k_D of at least 2 has been predicted for a B-like transition state.

c. It could be argued that the unsymmetrical transition state is the result of the symmetry of the alkene and would not be observed for symmetrical alkenes. The addition of p-methoxy groups to stilbene causes an equal rate enhancement of slightly more than four by each group. This is the expected result for statistical and unsymmetrical addition to either carbon of stilbene, without regard to substitution. The absence of an isotope effect (k_H/k_D = 0.98) for p-nitrostyrene is even more convincing. The decrease in rate shows that the addition is sensitive to the electron density of the π bond. The decreased electron density at the α carbon should have made for a more symmetrical transition state and therefore brought about a palpable secondary deuterium isotope effect at the α position. The observed result is still consistent with hybridization changes at only one end of the alkene.

Reference: R. P. Hanzlik and G. O. Shearer, J. Am. Chem. Soc., 97, 5231 (1975).

4. a. The large k_H/k_D values for A and B appear to be at least in part primary. Therefore, a hydrogen shift should be occurring in the transition state, and this mechanism would be termed k_Δ. The isotope effects for C and D are essentially normal for secondary systems that solvolyze without hydrogen shift. The faster rates for A and B are consistent with this interpretation, as are the product studies. Compounds A and B give a larger amount of rearranged

material, with the oxy functionality moved to the carbon bearing the methyl group. Compounds C and D on the other hand give a larger amount of unrearranged material, with the configuration inverted at the 1 position, as would be expected for direct solvent displacement ($\underline{k}_s$). All four substrates give about the same amount of elimination (73-87% tert-butylmethylcyclohexenes). Thus the isotope effects are not an accident of varying proportions of elimination vs. substitution. (See problem 7-7.)

b. Participation is always best in the antiperiplanar geometry, as exists between H and OTs in the ground state for A.

Because C and D have the H and OTs cis to each other, participation ($\underline{k}_\Delta$) from this geometry is impossible, and the slower $\underline{k}_s$ pathway becomes dominant.

c. Since B does not have an antiperiplanar relationship between H and OTs in the ground state, the molecule must convert to another conformation before H participation can occur. Whereas such a process is impossible for C and D (H (D) and OTs are cis), B can go into the ring-reversed chair (E) or into a twist boat (F) with antiperiplanar H and OTs. The chair (E) is probably unacceptable because of the axial

tert-butyl group. Therefore the most likely reactive conformation is the twist boat F, which avoids the axial tert-butyl group.

References: M. Pánková, J. Sicher, M. Tichý, and M. C. Whiting, J. Chem. Soc. B, 365 (1968); M. Tichý, J. Hapala, and J. Sicher, Tetrahedron Lett., 3739 (1969).

5. a. The reaction is a simple deprotonation/protonation, subject to general base catalysis. The large solvent isotope effect

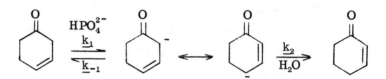

(7.7) is primary and indicates that the H—O (D—O) bond is being broken in the rate-determining step. In the above mechanism, the first step involves a C—H cleavage to form the charge-delocalized anion, and the second step involves O—H (O—D) cleavage to reprotonate the anion and form the iso-merized product. Thus the second step (k_2) must be rate determining. The anion can be protonated either at the 4 position to give the product (k_2) or at the 2 position to regenerate the starting material (k_{-1}). The latter process is revealed by incorporation of deuterium into the starting material. The observation that exchange is 575 times faster than isomerization $(k_{-1} \gg k_2)$ is consonant with a rate-determining second step. The rate of appearance of product is given by

$$\frac{d[P]}{dt} = k_2[S^-][H_2PO_4^-],$$

in which S^- is the substrate anion and $H_2PO_4^-$ is the protonated form of the general base (HPO_4^{2-}). Steady state in the anion (without explicitly including $[H_2O]$) gives

$$\frac{d[P]}{dt} = \frac{k_1 k_2 [S][HPO_4^{2-}]}{k_{-1} + k_2}$$

When $k_{-1} \gg k_2$, as in the present case $(K_1 = k_1/k_{-1})$,

$$\frac{d[P]}{dt} = K_1 k_2 [S][HPO_4^{2-}].$$

Additional terms would be needed for other general bases.
For cyclopentenone, there is essentially no solvent isotope

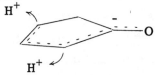

effect ($k_H/k_D = 0.9$). Consequently, the first step, cleavage
of a C—H bond by HPO_4^{2-} must at least in part be rate determining. The exchange (k_{-1}) and isomerization (k_2) are of
comparable rates, so the full kinetic expression must be
used. The rate in DPO_4^{2-}/D_2O was not observed to be uni-
formly first order. As deuterium is slowly incorporated into
the α position (k_{-1}), a larger proportion of C—D bonds are
being broken in the rate-determining step. Therefore the
rate is depressed by a primary isotope effect as the reaction
proceeds. The k_H/k_D of 0.9 was obtained from the first
4% of the reaction.

b. The diene structure is a useful model for the charge-delocalized anion intermediate. The planar cyclopentenone
anion allows protonation easily at both the α and the γ positions, so that the exchange (k_{-1}) and isomerization (k_2) rates

are comparable. For the cyclohexenone anion, the π overlap
between C_2 and C_3 is decreased by the nonplanarity of the
ring. To the extent that this overlap is decreased, protonation is less likely to occur at C_4. In the extreme of $\theta = 90°$

overlap is completely lacking between C_2 and C_3, so protonation at C_4 to form the C_2-C_3 double bond would be very unfavorable. At the other extreme ($\theta = 0°$), protonation at either C_2 (to give the C_3-C_4 double bond) or at C_4 (to give the C_2-C_3 double bond) would be favorable, as is the case for cyclopentenone. For cyclohexenone, the situation is intermediate ($\theta = 18°$). The decreased C_2-C_3 overlap makes C_4 protonation less likely and slows $\underline{k}_2$ with respect to $\underline{k}_{-1}$. Consequently, the second step becomes rate determining. There is no way to know whether the geometry of the anion parallels that of the diene, so that this argument may be fallacious. Barring other explanations, however, the difference in C_2-C_3 overlap caused by ring torsional constraints provides a reasonable explanation for the observed kinetic differences between the cyclohexenones and the cyclopentenones.

Reference: D. L. Whalen, J. F. Weimaster, A. M. Ross, and R. Radhe, J. Am. Chem. Soc., 98, 7319 (1976).

6. a. (i) Isotope effect. The absence of a primary isotope effect in A indicates that the first step is probably rate determining ($\underline{k}_{-1} \ll \underline{k}_2$). The large k_H/k_D in B requires that the second step be rate determining ($\underline{k}_2 \ll \underline{k}_{-1}$).

    (ii) Rate. The electronic effect of two sulfonate groups in B located two and four positions away, respectively, is larger than that of one group in A located two positions away. The standard deactivating effect of sulfonate therefore is larger in B than in A. This effect serves primarily to slow down the first step. In B, one of the sulfonate groups is peri to the electronically preferred site of substitution; A has no such group. The peri substituent should have a large effect on the second step (k_2) but little effect on the reverse of the first step (k_{-1}). When the proton leaves, the bulky electrophile (Ar-N=$\overset{+}{\text{N}}$:) must come up into the plane of the naphthalene ring, where it has the maximal interference with the peri sulfonate group. This steric effect slows down the second step (k_2) to the point that it becomes rate determining.

(iii) <u>Catalysis.</u> Pyridine assists in the removal of the proton in the second state. Since the first step is rate determining for A, addition of pyridine has no effect. Catalysis is observed for B, since proton removal is rate determining.

b. An increase in the rate of the second step for B through pyridine catalysis causes k_2 to become more nearly comparable to k_{-1}. The second step is no longer clearly rate determining. To the extent that the first step becomes kinetically important, the value of k_H/k_D is reduced. This system provides an example of catalysis that compensates for a steric effect.

Reference: F. Snyckers and H. Zollinger, <u>Helv</u>. <u>Chim</u>. <u>Acta</u>, <u>53</u>, 1294 (1970), and earlier papers by the same group cited in this source.

7. a. Unambiguous interpretation of a single overall k_H/k_D requires that the product-forming steps diverge after a common rate-determining step, such as formation of an ion pair. Dissection of the isotope effect into its components suggests that this approach is incorrect. The use of k_{iH}/k_{iD} requires that the rate-limiting and product-forming step be the same for a given product and be different from those of all other products. The various pathways could at most have a

$$R\text{—OBs} \xrightarrow{\text{slow}} R^{+-}\text{OBs} \begin{array}{c} \nearrow \\ \longrightarrow \\ \searrow \end{array} \begin{array}{l} H_2O \text{ substitution} \\ C_2H_5OH \text{ substitution} \\ \text{Elimination} \end{array}$$

common preequilibrium step. The substitution processes have a k_{iH}/k_{iD} of about 1.10, the elimination processes of about 1.18. Different degrees of bonding in the transition state are suggested. Since the overall k_H/k_D is a weighted average of the k_{iH}/k_{iD}, the exact overall value is an accident of the ratio of substitution to elimination (and of both these to rearrangement in other systems). Unless a solvolysis reaction leads to a single product, mechanistic interpretation of the overall k_H/k_D is not entirely reliable, except in a very qualitative fashion. The low value of k_{iH}/k_{iD} for the substitution products (1.10) suggests a fair degree of backside assistance by both components of the solvent in their respective transition states.

b. Each of the dissected isotope effects exhibits the normal decrease with temperature. The constancy of the overall k_H/k_D therefore has no mechanistic meaning at all. Any given k_H/k_D is the weighted average of all its components. Although each component k_{iH}/k_{iD} decreases with temperature, the overall average remains constant because of a greater weighting of the elimination components with their larger k_{iH}/k_{iD}. The abnormal overall isotope effect is the sum of a set of normal components. A caution therefore is in order to avoid interpreting the temperature dependence of secondary isotope effects, unless only one product-forming path exists.

References: G. A. Gregoriou and F. S. Varveri, Tetrahedron Lett., 287, 291 (1978).

8. a. The isotope effects for compound B correspond to the cyclo-addition reaction alone. The effect at the β position (0.897) is inverse and suggests considerable rehybridization from sp^2 to sp^3. On the other hand, the effect at the α position (0.994)

is negligible. Little rehybridization has occurred. Either
a stepwise mechanism (bond formation only at the β position)
or a very unsymmetrical concerted mechanism is possible.
Other cycloadditions have also exhibited very unequal isotope
effects, e.g., azodicarboxylates with enol ethers and
diphenylketene with styrenes. The isotope effects for compound A represent a mixture of the cycloaddition and ene
pathways. Since the ene reaction dominates the product (75%),
the α and β position isotope effects should for the most part
reflect the properties of the ene transition state. Consequently, the similarity of the figures for compounds A and
B is noteworthy. The ene reaction also must show considerable rehybridization from sp^2 to sp^3 in the transition state,
despite the fact that the β carbon is sp^2 in both starting
material and product. These results for A and B can be
explained in terms of two mechanisms, independent concerted
reactions or reactions via a common intermediate. For the
former case, the cycloaddition transition state must be
highly unsymmetrical (large β position effect, small α position effect) and the ene transition state must not lie on a
direct path from reactants to products (large β effect). For
the latter case, the simplest intermediate would be a perepoxide. In this mechanism, product formation occurs after

a common transition state. To satisfy the inequality of the
α and β positional effects, the bonds to oxygen would be
formed to different extents in the transition state leading to
the perepoxide. The anomalous isotope effect at the β
position for ene formation is then readily explained by the

rehybridization required to form the intermediate.

b. Because compound B lacks a γ proton, there is no basis to assess the relative size of the γ position isotope effect from the data given in (a) for the cycloaddition and the ene reaction. The γ position isotope effect, however, can be dissected into the cycloaddition and ene components. The independent concerted mechanisms are hard-pressed to explain the observed numbers. The figure of 1.376 (equivalent to a $\underline{k}_H/\underline{k}_D$ of 1.248) for the ene reaction is small for the primary isotope effect required of the independent concerted mechanism but not entirely out of line with various results observed by other workers in analogous cases. The low number would be consonant with an early transition state. The figure of 1.220 (equivalent to a $\underline{k}_H/\underline{k}_D$ of 1.148) for cycloaddition is even more bothersome in the independent concerted mechanism, since the γ hydrogen should not be involved at all in the cycloaddition transition state. An unusual, non-least-motion pathway involving some type of $O-H_\gamma$ interaction would have to be invoked. The perepoxide formulation more readily conforms to the data, since an interaction between H_γ and the incoming oxygen atoms is quite reasonable. The effect would be expected for both

products. Strong bonding at the β position and weak bonding at the α position permits some resonance interaction with the oxygen atom. Thus all data point toward a peroxide-like intermediate. The major flaw in this mechanism is the fact that the two dissected γ position isotope effects are not equal, as would have been expected for a common transition state. The independent concerted model has more serious flaws though, since the cycloaddition transition state would have to have an unusual interaction at the γ position and the ene transition state would have to have bonding primarily at the wrong position (β rather than γ).

Reference: A. A. Frimer, P. D. Bartlett, A. F. Boschung, and J. G. Jewett, J. Am. Chem. Soc., 99, 7977 (1977).

9. a. The active substrate below $[H_3O^+]$ = 0.03 M must be a protonated form of azulene-1-carboxylic acid, AzH^+CO_2H. For various reasons, the authors favored protonation at the 1 position. The independence of the rate on $[H_3O^+]$ at higher acidities suggests that a deprotonation step also is kinetically important. The species AzH^+CO_2H therefore probably converts to $AzH^+CO_2^-$ prior to decarboxylation. At $[H_3O^+]$ above 6 M, the carboxyl group does not deprotonate, so decarboxylation fails to occur. The authors point out that, at these high acidities, there is an equilibrium between the 1-protonated and the 3-protonated species, with the latter favored. The 3-protonated species, however, does not enter into the decarboxylation process. Protonation at the 1 position must activate the carboxyl group to deprotonation to the anion that can undergo loss of CO_2.

$$AzCO_2H + H_3O^+ \underset{k_{-1}}{\overset{k_1}{\rightleftharpoons}} AzH^+CO_2H + H_2O$$

$$AzH^+CO_2H + H_2O \underset{k_{-2}}{\overset{k_2}{\rightleftharpoons}} AzH^+CO_2^- + H_3O^+$$

$$AzH^+CO_2^- \xrightarrow{k_3} AzH + CO_2$$

The rate of formation of product is given by

$$\frac{d[AzH]}{dt} = k_3[AzH^+CO_2^-].$$

The steady state approximation in $[AzH^+CO_2^-]$ gives

$$k_2[AzH^+CO_2H] = k_{-2}[AzH^+CO_2^-][H_3O^+] + k_3[AzH^+CO_2^-]$$

or

$$[AzH^+CO_2^-] = \frac{k_2[AzH^+CO_2H]}{k_3 + k_{-2}[H_3O^+]},$$

so

$$\frac{d[AzH]}{dt} = \frac{k_2 k_3 [AzH^+CO_2H]}{k_3 + \underline{k}_{-2}[H_3O^+]} \ .$$

The steady state approximation in $[AzH^+CO_2H]$ gives

$$\underline{k}_1 [AzCO_2H][H_3O^+] + \underline{k}_{-2}[AzH^+CO_2^-][H_3O^+] =$$

$$\underline{k}_{-1}[AzH^+CO_2H] + \underline{k}_2[AzH^+CO_2H].$$

Substitution for $[AzH^+CO_2^-]$ and solution for $[AzH^+CO_2H]$ gives

$$[AzH^+CO_2H] = \frac{\underline{k}_1 [AzCO_2H][H_3O^+](\underline{k}_3 + \underline{k}_{-2}[H_3O^+])}{(\underline{k}_{-1} + \underline{k}_2)(\underline{k}_3 + \underline{k}_{-2}[H_3O^+]) - \underline{k}_2\underline{k}_{-2}[H_3O^+]} \ .$$

Simplification of the denominator and substitution into the rate expression gives

$$\frac{d[AzH]}{dt} = \frac{\underline{k}_1 \underline{k}_2 \underline{k}_3 [AzCO_2H][H_3O^+]}{\underline{k}_3(\underline{k}_{-1} + \underline{k}_2) + \underline{k}_{-1}\underline{k}_{-2}[H_3O^+]} \ .$$

Since proton loss is slower from carbon than from oxygen $(\underline{k}_2 \gg \underline{k}_{-1})$, the expression simplifies finally to

$$\frac{d[AzH]}{dt} = \frac{\underline{k}_1 \underline{k}_2 \underline{k}_3 [AzCO_2H][H_3O^+]}{\underline{k}_2 \underline{k}_3 + \underline{k}_{-1}\underline{k}_{-2}[H_3O^+]} \ .$$

At low acidity $(\underline{k}_2\underline{k}_3 \gg \underline{k}_{-1}\underline{k}_{-2}[H_3O^+])$, the rate law becomes

$$\frac{d[AzH]}{dt} = \underline{k}_1 [AzCO_2H][H_3O^+],$$

in agreement with the observed first order dependence on $[H_3O^+]$. At higher acidity $(\underline{k}_2\underline{k}_3 \ll \underline{k}_{-1}\underline{k}_{-2}[H_3O^+])$, the rate law becomes

$$\frac{d[AzH]}{dt} = \frac{\underline{k}_1 \underline{k}_2 \underline{k}_3}{\underline{k}_{-1}\underline{k}_{-2}} [AzCO_2H] = \underline{K}_1\underline{K}_2\underline{k}_3[AzCO_2H],$$

in agreement with the observed independence with respect to $[H_3O^+]$.

b. The above kinetic analysis suggests that carbon-carbon bond cleavage $(\underline{k}_3)$ is rate determining at intermediate acidities. The observed "full" $\underline{k}_{12}/\underline{k}_{13}$ of 1.039 is consistent with this conclusion. The lack of a heavy atom isotope effect at low acidities suggests that another step has become rate determining. Under these conditions, all evidence points toward a rate-determining protonation of the 1 carbon (the first step). Thus the carbon isotope effects are in full agreement with the acidity dependence.

References: J. L. Longridge and F.A. Long, J. Am. Chem. Soc., 90, 3092 (1968); H. H. Huang and F.A. Long, ibid., 91, 2872 (1969).

ALSO SEE PROBLEMS 8-7, 9-1, 9-7, 10-3, 10-4, 10-5, 10-6, 10-7, 10-10.

8

KINETIC SOLVENT EFFECTS

PROBLEMS

1. What is the effect on the rate of the following reactions from an increase in (a) the solvent dielectric constant (ϵ) and (b) the ionic strength (μ)?

(i) $C_6H_5S^- + C_6H_5CH_2\overset{+}{N}(CH_3)_3 \longrightarrow C_6H_5SCH_2C_6H_5 + N(CH_3)_3$

(ii) 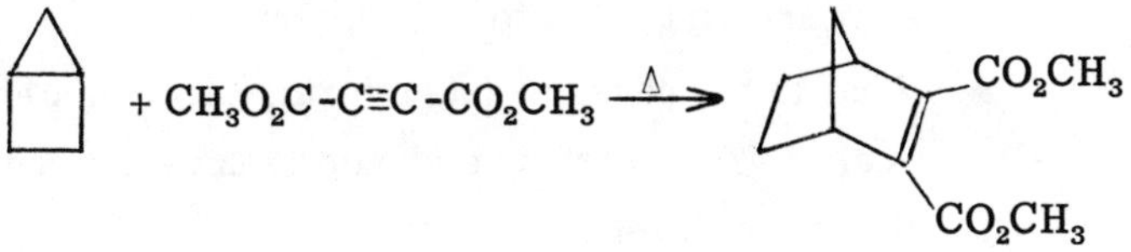 $+ CH_3O_2C-C{\equiv}C-CO_2CH_3 \overset{\Delta}{\longrightarrow}$

(iii) $+ {}^-OH \longrightarrow$

(iv) $\quad CH_3CO_2C_2H_5 + {}^-OH \longrightarrow CH_3CO_2H + {}^-OC_2H_5$

(v) $\quad (CH_3CH_2)_3N + CH_3CH_2I \longrightarrow (CH_3CH_2)_4N^+ I^-$

2. The hydrolysis of 2,4,6-trimethylbenzoyl chloride (A) is accelerated by the addition of lithium chloride, whereas added salt has little effect on the rate of p-nitrobenzoyl chloride (B). Explain.

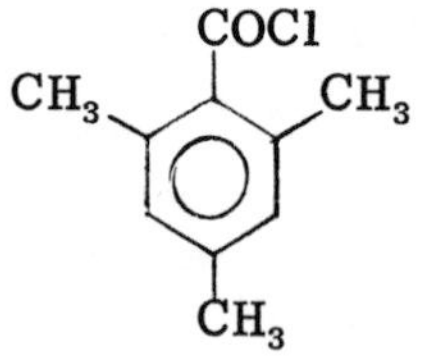

A 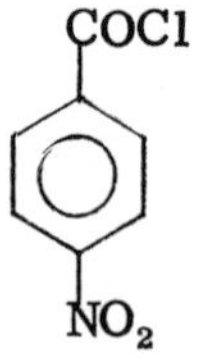   B

3. Berson and co-workers have devised a solvent polarity scale based on the Diels-Alder reaction between cyclopentadiene and methyl acrylate. Although the overall rate is not very

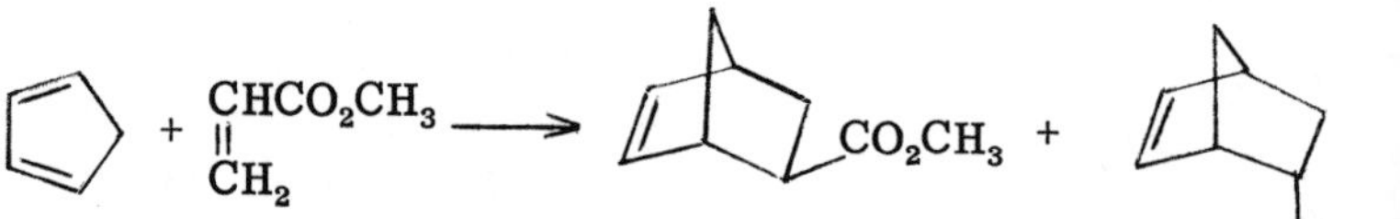

X   N CO_2C

sensitive to the solvent, the ratio of endo (N) to exo (X) products provides a useful scale. The solvent parameter Ω varies from

$$\Omega = \log \frac{[N]}{[X]} = \log \frac{k_N}{k_X}$$

0.44 for triethylamine to 0.84 for methanol.

a. What is the physical (structural, electronic) basis for this scale? Your response should include a discussion of transition state differences.

b. Should Ω correlate well with other solvent parameters such as $\underline{Y}$, $\underline{Z}$, and $\underline{E}_T$? Why?

4. Consider the following rate-solvent relationship, in which ϵ is the

$$\log \underline{k} = \frac{a}{\epsilon} + \underline{b}\,\underline{Q}_m + \underline{c}$$

solvent dielectric constant and $\underline{Q}_m$ is the heat of mixing of an aprotic solvent with 50 mol % $CHCl_3$ at 25 °C. For $CHCl_3$, $\underline{Q}_m = 0.0$ and $\epsilon = 4.8$; for 1,4-dioxane, $\underline{Q}_m = 453$ and $\epsilon = 2.2$; for acetone, $\underline{Q}_m = 420$ and $\epsilon = 20.7$; for DMSO, $\underline{Q}_m = 657$ and $\epsilon = 46.6$.

a. What solvent property does $\underline{Q}_m$ measure? What does ϵ measure?

b. For the reaction of $C_6H_5CO_2H$ with $(C_6H_5)_2C=N_2$, $\underline{a} = -1.80$, $\underline{b} = -0.0037$, and $\underline{c} = 1.12$. What two general solvent properties accelerate this reaction? Explain why in terms of the given mechanism.

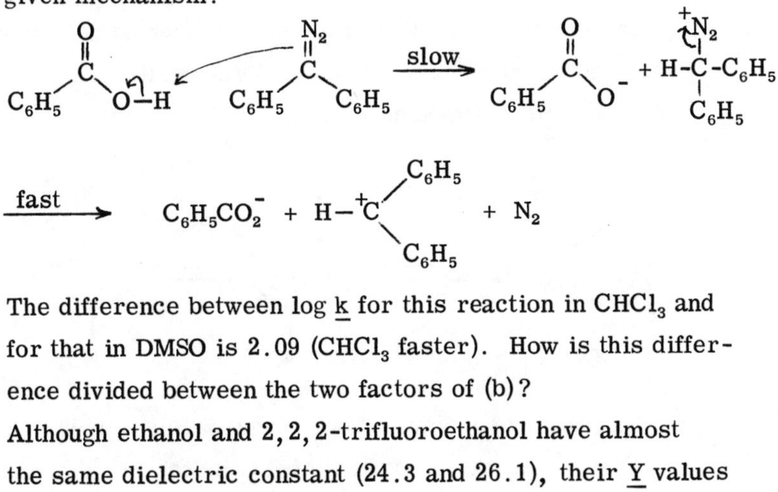

c. The difference between $\log \underline{k}$ for this reaction in $CHCl_3$ and for that in DMSO is 2.09 ($CHCl_3$ faster). How is this difference divided between the two factors of (b)?

5. a. Although ethanol and 2,2,2-trifluoroethanol have almost the same dielectric constant (24.3 and 26.1), their $\underline{Y}$ values (see problem 8-3b, answer) are quite different (-2.033 and +1.147). What are the reasons for these observations?

b. The Grunwald-Winstein plot (given on p. 142) of $\underline{k}_{solvent}/\underline{k}_{C_2H_5OH}$ vs. $\underline{Y}$ (mixed C_2H_5OH/CH_3CF_2OH at 85 °C) for the solvolysis of benzyl halides does not give a straight line. Why are the plots nonlinear? In the figure, the right-hand ordinate is for $ArCH_2Cl$, the left-hand is for $ArCH_2Br$.

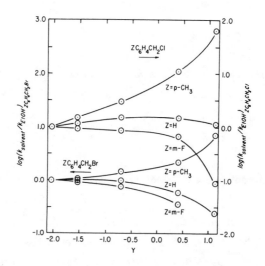

c. Why does the rate of the p-CH₃ compound increase with increasing Y but that of the m-F compound decrease ?

6. The reaction of sodium azide with 4-fluoronitrobenzene produces a nucleophilic substitution.

a. The reaction is carried out in aqueous dimethylformamide
(DMF). As the proportion of H_2O increases, the rate of
disappearance of reactants decreases drastically. Why?

b. In pure DMF, the rate is cleanly first order in each reactant,
but no NaF precipitates (NaF is very insoluble in DMF). If
5% H_2O is added to the solution of DMF/substrate/NaN₃,
NaF precipitates immediately. Explain these observations.

7. Related subject: kinetic isotope effects

Consider the following nucleophilic displacement.

$$C_6H_5S^- + C_6H_5CH_2\overset{+}{N}(CH_3)_2C_6H_5 \xrightarrow[0\ °C]{DMF} C_6H_5SCH_2C_6H_5 + (CH_3)_2NC_6H_5$$

a. The mechanism can be either S_N1 or S_N2. Which mechanism would dimethylformamide (DMF) favor and why? What then should be the kinetic order with respect to thiophenoxide ion and ammonium salt in DMF?

b. The $^{15}N/^{14}N$ isotope effect is 1.0200. Is this result in agreement with your conclusions in (a)? Explain.

c. Comparison of the reaction rates of $C_6H_5CH_2\overset{+}{N}(CH_3)_2C_6H_5$ and $C_6H_5CD_2\overset{+}{N}(CH_3)_2C_6H_5$ gives a $\underline{k}_H/\underline{k}_D$ of 1.19 (N.B., there are two α deuteriums). The normal S_N2 range (per deuterium) for secondary deuterium isotope effects is 0.95-1.04, and the normal S_N1 range is 1.07-1.25, depending on the leaving group. In consideration of your answers to (a) and (b), what can be said about $\underline{k}_H/\underline{k}_D$?

8. a. Taft and co-workers have suggested a six parameter, three term equation to describe solvent properties.

$$X = X_0 + \underline{a}\alpha + \underline{b}\beta + \underline{s}\pi *$$

In this equation, X is a reaction rate, equilibrium constant, or spectral property, X_0 is the property in the standard solvent, $\underline{a}$ describes the sensitivity to solvent acidities (hydrogen bond donor), $\underline{b}$ describes the sensitivity to solvent basicities (hydrogen bond acceptor), and $\underline{s}$ describes the sensitivity to solvent polarity and polarizability. The α, β, and $\pi *$ scales were based on the solvatochromic properties of various primary indicators. Comment on the form of this equation.

b. Plots of $\pi *$ vs. various other empirical solvent scales, e.g., Dimroth's $\underline{E}_T$, Brooker's χ_R, Brownstein's adaptation $\underline{S}$ of Kosower's $\underline{Z}$, and others, give very scattered plots. The elimination of hydrogen-bonding (CH_3OH, H_2O, CF_3CH_2OH, etc.), halogenated (CH_2Cl_2, CCl_4, $Cl_2C=CHCl$, etc.) or highly polarizable (CS_2, benzene, anisole, nitrobenzene, etc.) solvents give satisfactory ($\underline{r} > 0.96$) plots of $\pi *$ vs. all these other parameters, as well as vs. the solvent dipole moment

μ. The dielectric function $(\epsilon - 1)/(2\epsilon + 1)$ gives only an average correlation with π^* ($\underline{r} = 0.91$). What is the meaning of these results, particularly the observations concerning μ and ϵ ?

SOLUTIONS

1. a. The solvent theory developed by Hughes and Ingold states
        that an increase in ε causes an increase in rate when the
        transition state is more polar than the ground state and
        causes a decrease in rate when it is less polar. Thus in
        many instances the effect of solvent ε on the rate can be pre-
        dicted simply from the relative polarity of the ground and
        transition states.

    (i)   In this displacement reaction, two oppositely charged
           species are brought together. The transition state
           should be less polar than the ground state, so an
           increase in ε will slow the reaction.

    (ii)   If the reaction is concerted or passes through a di-
           radical (A), a change in ε will have little effect on the
           rate. A dipolar intermediate (B) would be indicated

A   B

           by an increase in rate. The observed rate shows little
           solvent sensitivity (relative rates, C_6H_6 1.46,
           CH_3CH_2OAc 1.00, CH_3CN 1.85), so the dipolar species
           B can be eliminated.

    (iii) Here two ions of the same charge are brought together
           in a transition state that must be more polar than the
           ground state. The rate will increase in a solvent of
           higher ε.

    (iv)   In the basic ester hydrolysis reaction of a neutral sub-
           strate, the reacting species as a whole maintain a sin-
           gle negative charge throughout the reaction. The
           charge in the transition state (C), however, is more

C

$$CH_3 \overset{\displaystyle O^{\delta-}}{\underset{\displaystyle OC_2H_5}{\overset{\displaystyle \|}{\underset{}{\overset{\displaystyle C}{\diagdown}}}}} \overset{\delta-}{OH}$$

dispersed than in the starting hydroxyl group, so that less dipolar solvation is needed. Thus the rate decreases slightly with an increase in ϵ.

(v)   Here two polar, uncharged molecules combine in the Menschutkin transition state with an incipient separation of charge. The rate will increase with an increase in ϵ (relative rates, hexane 1.0, C_6H_6 82, acetone 840, nitrobenzene 2760).

b. The rate of an ionic reaction is proportional to the square root of the ionic strength. This equation applies to aqueous

$$\log \underline{k} = \log \underline{k}_0 + 1.018\, \underline{Z}_A \underline{Z}_B\, \mu^{\frac{1}{2}}$$

solutions at 25 °C, and $\underline{Z}_A$ and $\underline{Z}_B$ are the formal charges on the two reactants in a bimolecular reaction. The primary salt effect therefore depends only on the relative sign and the magnitude of the written charges. If $\underline{Z}_A$ and $\underline{Z}_B$ have the same sign (positive or negative), an increase in μ causes an increase in the rate. If they have the opposite sign, the rate decreases. If either species is neutral, there should be little or no salt effect. This approach works well for charged species, but salt effects can also be observed for neutral cases, particularly when charge is appreciably separated in the transition state.

(i)   The charges are unlike, so the rate decreases with μ.

(ii)   There should be little salt effect in this reaction between neutrals.

(iii)   Here $\underline{Z}_A = \underline{Z}_B = -1$, so there is a large positive salt effect (increased rate with μ).

146

(iv) The charge is zero on one of the reactants, so the salt effect is small.

(v) This reaction between neutrals should not be very sensitive to μ, but a positive salt effect might be expected because of the polarity of the ground and transition states.

References: E.S. Gould, "Mechanism and Structure in Organic Chemistry," Holt, Rinehart and Winston, New York, NY, 1959, pp 183-87; K. B. Wiberg, "Physical Organic Chemistry," Wiley, New York, NY, 1964, pp 374-95; the data for system (ii) are from the work of P. P. Gassman.

2.

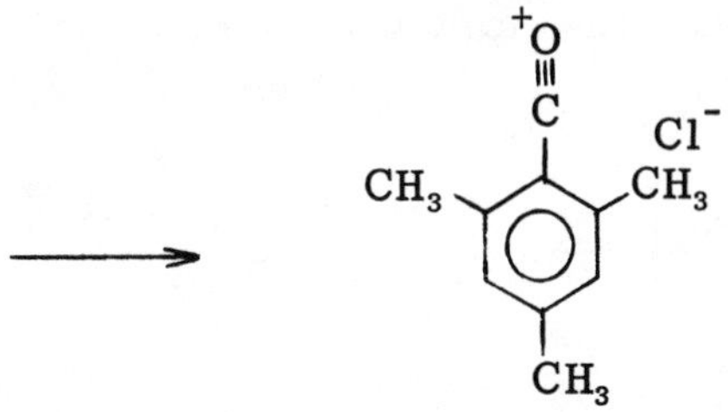

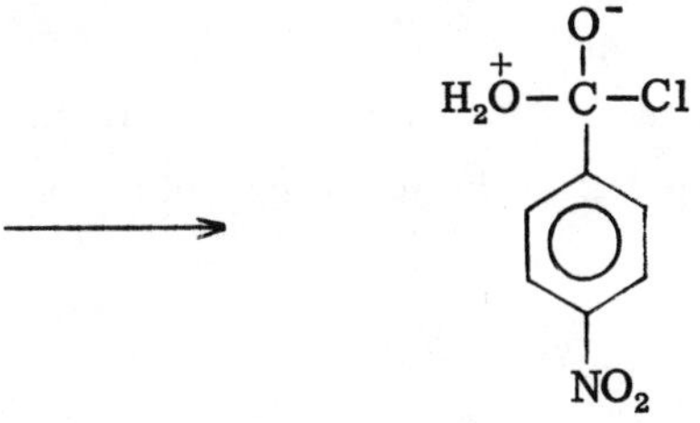

Both systems (A and B) involve the reaction of neutrals (see problem 8-1b), with separation of charge in the transition state. The mechanism for hydrolysis of A is relatively unusual. The ortho methyl groups prevent the formation of the usual tetrahedral intermediate. The only viable pathway is the unimolecular departure of the chloride to give an acylonium ion (C). The linear $C-C{\equiv}O^+$ group avoids the ortho methyl groups. This mechanism has also been observed in the hydrolysis of esters, but it is more common in the hydrolysis of acid chlorides because of the better leaving group properties of chloride ion. Acid chloride hydrolysis can occur readily under neutral conditions. In the transition state leading to intermediate C, there is considerable separation of charge, so the salt effect is large. The unhindered system B proceeds to the normal tetrahedral intermediate D, in which charge separation is small. As a result, there is little or no salt effect.

Reference: J. March, "Advanced Organic Chemistry," 2nd ed, McGraw-Hill, New York, NY, 1977, pp 333, 347-48.

3. a. The transition states must have slight differences in polarities, so that solvent interactions can influence the product distribution. The dipoles of the starting materials are as follows. The dipole in cyclopentadiene results from

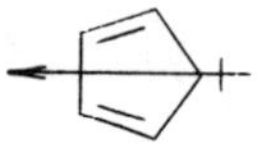

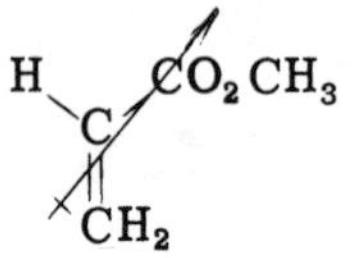

the higher electronegativity of sp² carbon atoms with respect to that of sp³ carbon atoms. The transition states combine their dipoles in the following manner.

Endo

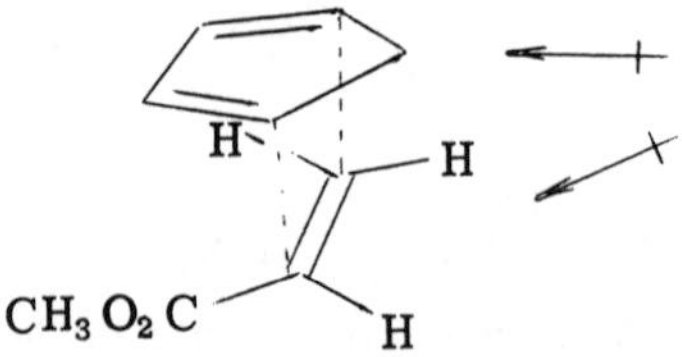

Exo

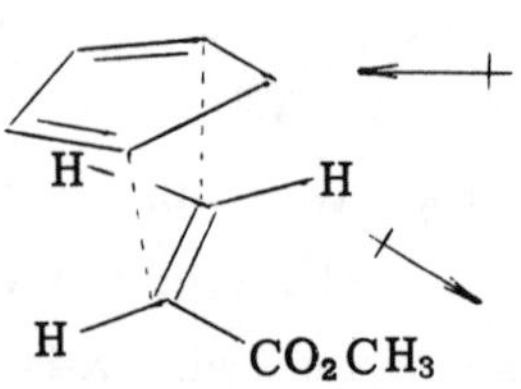

In the endo form the dipoles reinforce, whereas in the exo form they partially cancel. Therefore the endo transition state is more polar and is favored by more polar solvents, as observed.

b. The Grunwald-Winstein $\underline{Y}$ value is given by the equation,

$$\log \frac{\underline{k}}{\underline{k}_0} = \underline{m}\,\underline{Y}\,,$$

in which $\underline{k}$ is the rate in any solvent, $\underline{k}_0$ is the rate in 80% ethanol, $\underline{Y}$ is the solvent ionizing power (0.0 for 80% ethanol), and $\underline{m}$ is the sensitivity of the given system to ionizing power (1.00 for tert-butyl chloride at 25 °C). The Kosower $\underline{Z}$ value is the energy in kcal mol^{-1} of the longest wavelength band for system A. The Dimroth $\underline{E}_T$ value measures the analogous band for B. For $\underline{Z}$ and $\underline{E}_T$, the numbers pertain to the charge-transfer process ($M^+N^- \longrightarrow M \cdot N \cdot$). All of these

A   B

processes--tert-butyl chloride solvolysis, charge transfer, Diels-Alder reaction--alter charge separation within the total system. Charge separation may either increase (solvolysis, endo Diels-Alder) or decrease (charge transfer, exo Diels-Alder). Therefore a similar dependence of the measurables (Ω, $\underline{Y}$, $\underline{Z}$, $\underline{E}_T$) should result, but the sign of the slope of the rate relationship is not always the same. The choice among these solvent parameters depends on the ease of measurement (solvolysis or Diels-Alder rates take longer to measure than solvatochromic differences in electronic transitions) or the

range of solvents (solvolysis or Diels-Alder reactions have
a limited selection). Dimroth's $\underline{E}_T$ value is enjoying current
popularity because of its availability for a wide range of
solvents, but $\underline{Y}$ is used almost universally in solvolysis
studies. Some subtle differences do exist between the various parameters, and the final choice may depend on the
specific type of reaction under consideration.

References: J. A. Berson, Z. Hamlet, and W. A. Mueller,
J. Am. Chem. Soc., 84, 297 (1962); J. A. Hirsch, "Concepts
in Theoretical Organic Chemistry," Allyn and Bacon, Boston,
MA, 1974, pp 188-91, 208-14.

4. a. The dielectric constant is a measure of general electro-
    static interactions (polarity and polarizability) of the solvent.
    The quantity $\underline{Q}_m$ measures any specific interaction of a given
    solvent with $C\overline{H}Cl_3$. Since the dominant interaction provided
    by $CHCl_3$ (other than that measured by ε) is hydrogen bonding,
    $\underline{Q}_m$ is a measure of the receptivity of the solvent for a hydro-
    gen bond, i.e., solvent basicity.

    b. Since $\underline{a}$ is negative and ε is in the denominator, the rate is
    accelerated by an increase in general solvent electrostatic
    interactions. This result follows from the fact that the tran-
    sition state is more polar than the ground state (problem
    8-1a, type (v)). Since $\underline{b}$ is negative and $\underline{Q}_m$ is in the numera-
    tor, the rate is accelerated by a decrease in solvent basicity.
    Do not be misled by the small value of $\underline{b}$, since the typical
    values of $\underline{Q}_m$ are quite large. A highly basic solvent sta-
    bilizes the ground state through hydrogen bonding (C_6H_5-CO-
    O-H---:S), so the transition state is more difficult to attain.
    Thus either high polarity (ε) or low basicity ($\underline{Q}_m$) accelerates
    the reaction.

    c. Subtraction of the defining equation for two solvents ($CHCl_3$
    is labeled 1, DMSO 2) gives

$$\Delta \log \underline{k} = 1.8 \left(\frac{1}{\epsilon_2} - \frac{1}{\epsilon_1} \right) - 0.0037 \left(\underline{Q}_{m1} - \underline{Q}_{m2} \right).$$

The first term is -0.34, the second (including the minus sign) is 2.43, and the sum as given is 2.09. Thus the rate is dominated (86%) by the basicity term. In this hydrogen transfer reaction, destabilization of the ground state through low solvent basicity is much more important than stabilization of the transition state through general electrostatic interactions. A cautionary note should be inserted. The defining equation contains four parameters ($\underline{a}$, ϵ, $\underline{b}$, $\underline{Q}_m$). Any such equation can appear to correlate reaction rates facilely, because of the large number of adjustable parameters. Although the quantitative success of the equation (the correlation coefficient for the above reaction in five solvents is 0.996) should not be overemphasized, the assessment of the approximate relative importance of dielectric and basicity factors is useful.

References: A. Buckley, N.B. Chapman, M.R.J. Dack, J. Shorter, and H.M. Wall, J. Chem. Soc. B, 631 (1968); M. R. J. Dack, Chemtech, 1 , 108 (1971).

5. a. Although the two solvents have similar abilities to interact with solutes through general electrostatic processes, as measured by ϵ, they exhibit considerably different ionizing powers, as measured by $\underline{Y}$. Whereas pure C_2H_5OH is less ionizing than the standard solvent (80% aqueous C_2H_5OH), CH_3CF_2OH is much more ionizing. Compared to pure CH_3CH_2OH ($\underline{Y} = -2.033$), the presence of 20% H_2O in aqueous CH_3CH_2OH ($\underline{Y} = 0.0$) must provide better cation/anion stabilization during the process of ionization. The ability of H_2O to stabilize both cations (X^+---OH_2) and anions (Y^----HOH) is well documented. The higher ionizing power of trifluoroethanol ($\underline{Y} = 1.147$) is shared with other fluorinated solvents, such as CF_3CO_2H and $(CF_3)_2CHOH$. The mechanism

of enhancement has never been fully clarified. Dannenberg has suggested that the added cation stabilization is due to direct interactions with the fluorine atoms, as in A. This model also explains the very low nucleophilicity of these

	X	Y
	CH_3	OH
	F	$CHCF_3OH$
	F	CO_2H

solvents. In H_2O, alcohols, and carboxylic acids (nonfluorinated), cation stabilization occurs via nucleophilic centers (OH, C=O). In these fluorinated solvents, however, the stabilizing C-F bonds are at the opposite end of the molecule from the nucleophilic functionalities.

b. To give a linear Grunwald-Winstein plot, either a reaction must depend almost entirely on the ionizing power of the solvent or other factors must be constant, as in the defining reactions, tert-butyl chloride or 2-adamantyl chloride solvolysis. The curvature of the plot in the present instance indicates that the single property of ionizing power (bond breaking) is insufficient and that nucleophilicity (bond making) must be taken into consideration as well. The four parameter equation includes the solvent nucleophilicity $\underline{N}$ and the

$$\log \underline{k}/\underline{k}_0 \ = \ \underline{m}\,\underline{Y} \ + \ \underline{\ell}\,\underline{N}$$

substrate sensitivity to nucleophilicity $\underline{\ell}$ as well as the ionizing power term. Use of this equation should produce a more nearly linear plot. The curvature also indicates that $\underline{Y}$ and $\underline{N}$ are not linearly (negative slope) related.

c. The $\underline{p}$-CH_3 group increases the limiting carbonium ion $(\underline{k}_C)$ character of the transition state (more bond breaking) through polar or hyperconjugative electron donation. As $\underline{Y}$ increases, the substrate is increasingly disposed toward bond breaking at the expense of nucleophilic assistance. The electron-withdrawing character of m-F both reduces the stability of the carbonium ion (k_C component) and increases the likelihood of S_N2 solvent displacement (k_S component, more bond making). Electron withdrawal makes the site more electrophilic and hence more susceptible to nucleophilic attack. The balance between Y and N sensitivity is clearly dependent on the electron demand at the reaction site.

References: D. A. da Roza, L. J. Andrews, and R. M. Keefer, J. Am. Chem. Soc., 95, 7003 (1973); J. J. Dannenberg, Angew. Chem., Int. Ed. Engl., 14, 641 (1975); J. Kaspi and Z. Rappoport, Tetrahedron Lett., 2035 (1977).

6. a. Whereas protic solvents such as H_2O and alcohols stabilize both cations and anions, many so-called aprotic, dipolar solvents (DMF, DMSO, SO_2, HMPA, acetone, sulfolane, acetonitrile) can effectively stabilize only the cation. In these solvents the negative end of the dipole (as in $CH_3 \overset{O}{\underset{+}{-S}} -CH_3$) is relatively exposed and able to solvate cations, whereas the positive end is buried in a polarizable hydrocarbon environment and is poorly accessible to anions. The poorer solvation of anions increases their reactivity. Aprotic, dipolar solvents destabilize the ground state, in which the anion is relatively localized (e.g., $^-N_3$), to a greater extent than they destabilize the transition state, in which the negative charge is more dispersed. Not only does dispersed negative charge require less solvation, but the dispersed positive charge of the solvent is better able to provide solvation. The lack of ground state solvation in DMF makes $^-N_3$ a better nucleophile. As H_2O is added, the more heavily solvated anion is a less effective nucleophile, and the reaction is slowed.

   b. Whereas the substitution intermediate is formed very readily in DMF for the reasons given in (a), it is not eager to decompose to the product because of the poor solvation of the fluoride ion that would be formed. Thus the intermediate is stable under the reaction conditions, trapped in a potential well with either $^-N_3$ or F^- on the other sides of the barriers. As H_2O is added, F^- is stabilized relative to the intermediate, so the σ complex collapses to product. The concentration of Na^+ and F^- exceeds that permitted by the ion product, so NaF precipitates. The results are summarized in the energy diagram.

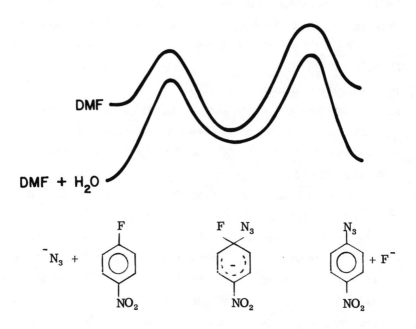

The left-hand side of the diagram shows that the ground state ($^-N_3$) is destabilized in DMF more than the transition state, so the activation energy is smaller in DMF and the rate is faster (see part (a)). The middle part shows that the intermediate in DMF is stabler than either the starting material or the product, whereas in DMF + H_2O, the intermediate is unstable with respect to the product.

Reference: E. M. Kosower, "An Introduction to Physical Organic Chemistry," Wiley, New York, NY, 1968, pp 334-37.

7. a. Because DMF solvates anions so poorly (see problem 8-6),
thiophenoxide should exhibit enhanced nucleophilic character.
The negative charge in the transition state (A) is more

$$A \quad C_6H_5S \overset{\delta-}{-}\!-\!-\!-\overset{\overset{\displaystyle C_6H_5}{|}}{\underset{\underset{\displaystyle H \quad H}{}}{C}}\!-\!-\!-\!-\overset{\delta+}{N}(CH_3)_2C_6H_5$$

dispersed, so the anionic destabilization is less than that of
thiophenoxide. Therefore, DMF should favor the S_N2 process (in comparison to the situation in a protic, hydroxylic
solvent), and the rate should be first order in both thiophenoxide and ammonium salt. Indeed, this order is observed.

b. The ratio of zero point energies and hence the ratio of reaction rates (k_A/k_B) is inversely proportional to the ratio of
reduced masses (μ_B/μ_A) (the reduced mass is given by

$$\mu_A = \frac{m_A m_X}{m_A + m_X} \quad ,$$

in which A and X are the atoms that form the bond). If the
reduced masses differ markably, as they do for C-H vs.
C-D, a sizable primary isotope can be expected. If they
differ only slightly (heavy atom isotope effects), as in the
present case of C-^{14}N vs. C-^{15}N, quite a small isotope effect
results. Thus a k_{15N}/k_{14N} of 1.0200 is a primary isotope
effect and reflects bond breaking in the transition state,
despite its small size. This isotope effect is consistent with
an S_N2 process with C-N bond breaking in the rate-determining step. It probably excludes a rapid preequilibrium
C-N cleavage followed by a rate-determining break-up of the
resultant ion pair.

c. The k_H/k_D of 1.19 corresponds to about 1.09 per deuterium
atom. This value is definitely out of the given S_N2 range
(≤ 1.04). The order of the reaction in DMF and the magnitude
of the nitrogen isotope effect, however, demand that the

mechanism be S_N2. The fault must lie in the previously quoted S_N2 range (0.95-1.04). Expansion to 1.09, however, causes a serious overlap of the S_N1 and S_N2 ranges and compromises the mechanistic criterion to a considerable extent. Why is the α secondary isotope effect so large for the present case? The usual cause of α effects is differences in the zero point energies of C_α-H and C_α-D out of plane bending vibrations. The very bulky leaving group ($N(CH_3)_2C_6H$ which is even larger than tert-butyl) places the C_α-H (C_α-D) bonds in a very crowded environment. The zero point energy differences between C_α-H and C_α-D therefore are larger than normal in the ground state. In the transition state, the large group (N, N-dimethylaniline) has moved away from the α carbon, so these enhanced steric differences are removed for the most part. Thus the difference in zero point energy differences (ground state to transition state) and hence k_H/k_D are larger than in an analogous, uncongested system. It can be concluded from this study that the α secondary deuterium isotope effect is not a valid criterion for distinguishing S_N1 and S_N2 mechanisms in borderline situations, particularly when the leaving group is especially large.

Reference: K. C. Westaway, Tetrahedron Lett., 4229 (1975).

8. a. By using a three term expression, these workers have tried, with considerable success, to isolate the specific solvent properties of importance. They have examined acidity, basicity, polarity, and polarizability and have developed scales for each factor. Thus α runs from 0.29 for CH_3CN up to 1.017 for H_2O, β from 0.247 for anisole up to 0.990 for hexamethylphosphoramide, and π^* from -0.081 for hexane up to 1.118 for formamide. This equation probably stands on firmer theoretical grounds than any other, simply because each solvent property is evaluated by a separate variable. Use of the α-β-π^* scale, however, requires the measurement of much larger data sets than previously, since up to three

adjustable parameters (a, b, s) must be determined. Small
data sets would not yield meaningful figures, unless it was
obvious that one or two of the factors were negligible.

b. The lack of correlation between π^*, which is free of any
dependence on hydrogen bonding, and the other empirical
solvent parameters indicates the variability of the hydrogen-
bonding factor in the other parameters. The elimination of
hydrogen-bonding solvents should reasonably give better
correlations. The s π^* term, however, contains the dependence on both the polarity and the polarizability of the solvents.
The authors find that highly polarizable solvents (aromatic,
polyhalogenated) also must be excluded from the plots before
reasonable correlations are achieved. After elimination of
hydrogen-bonding, aromatic, and polyhalogenated solvents,
π^* correlates well with all the various solvent parameters,
and the entire set in turn correlates well with the dipole
moment μ. For the aprotic, dipolar solvents that remain in
the set (ethyl ether, ethyl acetate, acetone, cyclohexanone,
nitromethane, dimethyl sulfoxide, etc.), the dominant
solvent-solute interaction must be dipolar, and the molecular
dipole moment represents the "elusive idealized solute-
independent solvent polarity parameter." The poor dependence of π^* (and thus all the other empirical solvent parameters) with the dielectric function $(\epsilon - 1)/(2\epsilon + 1)$ indicates
that specific solute-solvent dipolar interactions, rather than
a bulk effect of the solvent acting as a continuous dielectric,
provide the major solvent effect for these systems. The
exclusion of highly polarizable solvents from this treatment
suggests that yet another term must be added. The authors
have suggested replacing s π^* by s $(\pi^* + d\ \delta)$, in which $d = 0.0$
for the "select," nonpolarizable solvents, $d = 0.50$ for polyhalogenated solvents, and $d = 1.00$ for aromatic solvents.
An overlooked family of polarizable materials is those con-

taining triple bonds. Although the authors exclude CS_2 from their "select" set, they retain $CH_3C{\equiv}N$. Nonetheless, CH_3CN is excluded from their π^* vs. μ plot, because of inconsistencies between the measured value (0.713) and that expected from other parameters. Other functionalities, e.g., nitro and ester, may also have small contributions from polarizability (methyl formate, methyl acetate, and nitroethane have some of the largest deviations from the π^* vs. μ plot). Ultimately, values of $\underline{d}$ may have to be sought for all π-electron-containing solvents.

References: M. J. Kamlet and R.W. Taft, J. Am. Chem. Soc., 98, 377 (1976); R.W. Taft and M.J. Kamlet, ibid., 98, 2886 (1976); M. J. Kamlet, J. L. Abboud, and R.W. Taft, ibid., 99, 6027 (1977); J. L. Abboud, M. J. Kamlet, and R.W. Taft, ibid., 99, 8325 (1977).

ALSO SEE PROBLEMS 9-2, 9-5, 10-8, 10-12.

9

KINETIC ELECTRONIC
AND STERIC EFFECTS

PROBLEMS

1. Related subject: kinetic isotope effects

Electrophilic iodine (I^+) can be generated electrochemically in acetonitrile. Iodination of monosubstituted benzenes (Ar—X) in this manner gives a Hammett slope ρ (σ^+) of -6.27. Deuterium isotope effects (D para to X) (k_H/k_D) are in the range 1.5-4.5, depending on X.

a. Write down the overall mechanism, with benzene (X = H) as your example. Indicate which step appears to be rate determining and explain your conclusion in terms of ρ and k_H/k_D.

b. The k_H/k_D varies with X, OCH$_3$ 1.45, CH$_3$ 1.54, H 2.25, Cl 4.46, CO$_2$CH$_3$ 1.50. Give two explanations for the maximum at Cl (i.e., the decrease when X is either more or less electron withdrawing than Cl).

2. <u>Related subjects</u>: <u>introduction to kinetics, kinetic solvent effects</u>

The decarboxylation of 3-carboxybenzisoxazoles is thought to

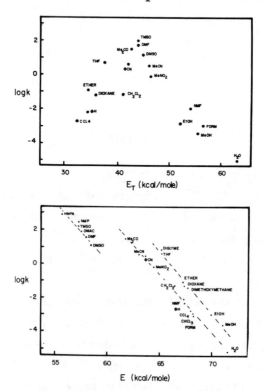

take place in the carboxylate (anionic) form in a concerted fashion, i.e., removal of CO_2 and ring opening occur at the same time. There is no catalysis by acid or base, and the neutral carboxylic acid is unreactive. The $\Delta S^{\ddagger}$ is constant (about +20 eu) in all solvents. The two figures below show the variation of log $\underline{k}$ with Dimroth's $\underline{E}_T$ (see problem 8-3) and with the

long wavelength maximum of the anion of 2-cyano-5-nitrophenol ($\underline{E}$).

a. Why is the log $\underline{k}$ correlation better with $\underline{E}$ than with $\underline{E}_T$?

160

b. Explain why the dipolar, aprotic solvents (DMF, DMSO, etc.) in general react several orders of magnitude more rapidly than the protic solvents.

c. Explain why H_2O is by far the slowest solvent.

d. In H_2O, $\rho = 1.37$; in HMPA (hexamethylphosphoramide), $\rho = 2.40$. Explain.

3. Related subject: <u>introduction to kinetics</u>

a. The electrophilic addition of hydrogen chloride to phenyl-allene in acetic acid gives only cinnamyl chloride. Correlation of the rates with σ^+ produces a linear Hammett plot with

$$ArCH{=}C{=}CH_2 \xrightarrow{\text{HCl}} ArCH{=}CHCH_2Cl$$

$\rho = -4.2$. The solvolyses of $ArC(CH_3)_2Cl$ and $ArCHCH_3Cl$ also have a ρ (σ^+) of about -4. Suggest the structure of the cationic intermediate (reason from the Hammond postulate).

b. Addition of hydrogen chloride in acetic acid to <u>trans</u>-1-phenyl-1,3-butadiene gives a ρ (σ^+) of -2.98, whereas ρ (σ^+) for 1-phenyl-1,2-butadiene is about -4.2 (part (a)). The two compounds give the same product, <u>trans</u>-1-methyl-3-phenyl-allyl chloride. Account for the decrease in ρ for the 1,3-diene.

4. A key step in the laboratory synthesis of estrone is a Lewis acid-catalyzed double cyclization. The Hammett plot is curved

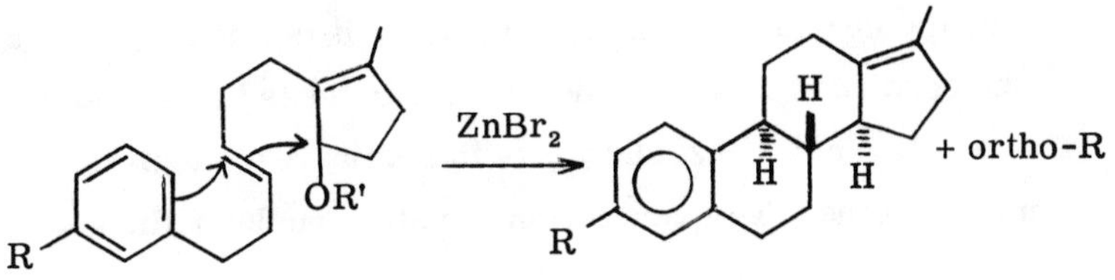

A   B (para-R)   C

concave downwards, with $\rho \sim 0$ for $\sigma < 0$ (R = CH_3O, CH_3) and $\rho \sim -3$ for $\sigma > 0$ (R = Cl, CF_3).

a. From the observation of a curved Hammett plot of this type, what can be said about the mechanism?

b. Explain why ρ is nearly zero for $\sigma < 0$ but large and negative for $\sigma > 0$.

c. The concentration ratio of para-R (B) to ortho-R (C) in the product depends on the nature of the leaving group (R' = H, $(CH_3)_3Si$, or C_6H_5CO). What can be said about the timing of closure of the two rings?

5. Related subject: kinetic solvent effects

The bromination of stilbene can occur either by initial addition to one end of the double bond to give an open carbonium ion (A) or by initial addition to the middle to form a cyclic bromonium ion (B).

$$ArCH=CHAr' \xrightarrow{Br_2} Ar\overset{+}{C}HCHBrAr' + ArCHBr\overset{+}{C}HAr' \quad (A)$$

$$ArCH=CHAr' \xrightarrow{Br_2} ArCH\overset{\overset{+}{Br}}{\diagdown} CHAr' \quad (B)$$

$$Ar = \underset{X}{\bigcirc} \qquad Ar' = \underset{Y}{\bigcirc}$$

a. What is the form of the Hammett equation for mechanism A (use the correct σ constants)?

b. What is the form of the Hammett equation for mechanism B?

c. In CH_3OH, the simple Hammett plot (log $\underline{k}$ vs. the sum of the σ constants) is curved. In CCl_4, it is linear. What can be concluded about the mechanism in these two solvents?

6. The reaction of a benzaldehyde with semicarbazide gives a semicarbazone. The mechanism involves nucleophilic addition

$$Ar-CHO + NH_2NH\overset{O}{\overset{\|}{C}}NH_2 \xrightarrow{H^+} Ar-CH=NNH\overset{O}{\overset{\|}{C}}NH_2$$

followed by elimination. The Hammett plot is pH dependent.

a. Write out the mechanism, showing only two overall steps.

b. Why is ρ fairly large (0.91) and positive for pH 1.75?

c. Why is ρ almost zero (0.07) for pH 7.0?

d. Why is there a break in the curve for pH 3.9?

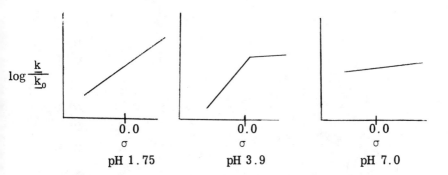

$\log \dfrac{k}{k_0}$

0.0	0.0	0.0
σ	σ	σ
pH 1.75	pH 3.9	pH 7.0

7. Related subjects: introduction to kinetics, kinetic isotope effects

The cyclodehydration of 2-phenyltriarylcarbinols in 80% acetic acid containing 4% sulfuric acid occurs by the following mechanism.

Substituents were placed on either or both Ar rings (A).

A

Cpd	X	Y	Z
1	Cl	H	Cl
2	H	CF_3	H
3	H	Cl	H
4	Cl	H	H
5	H	OCH_3	H
6	H	H	H
7	H	CH_3	H
8	CH_3	H	H
9	OCH_3	H	H
10	CH_3	H	CH_3
11	CH_3	H	OCH_3
12	OCH_3	H	OCH_3

The Hammett plot, in which the numbers correspond to the above compounds, shows a sharp break.

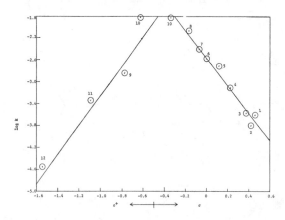

a. In terms of the mechanism, why is there a break in the Hammett plot, and why is the plot concave downwards?

b. Why is σ^+ used for the substituents on the left of the plot and σ for those on the right?

c. Why is ρ negative (-2.51) on the right but positive (2.77) on the left?

d. Which group of compounds (those on the left or those on the right) has the more negative entropy of activation? Why?

e. Which group has an acid dependent rate? Why?

f. Which group shows a rate acceleration in D_2O, compared to H_2O? Why?

8. The methoxide-catalyzed methanolysis of aliphatic menthyl esters has the following mechanism. For nine aliphatic sub-stituents $R[(CH_3)_3C, (CH_3)_3CCH_2, C_2H_5, CH_3, H, C_6H_5OCH_2,$

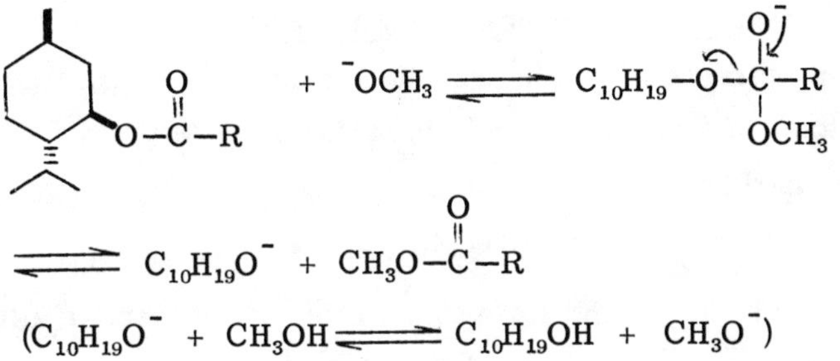

$ClCH_2, NCCH_2,$ and $Cl_2CH]$, the rate follows the two term Taft equation, with $\rho^* = 2.70 \pm 0.07$ and $\delta = 1.30 \pm 0.06$.

a. What is the significance of the magnitude of δ, in comparison with the Taft model reaction? What is the cause of any differences?

b. For the same reaction, except with meta- and para-substi-tuted phenyl rings for R, the Hammett plot gives $\rho = 2.63$. What is the significance of the identity of ρ and ρ^*?

9. Charton has proposed a steric substituent parameter $\underline{v}_X = \underline{r}_X - 1.20$, in which $\underline{r}_X$ is the van der Waals radius of a group X and 1.20 is the van der Waals radius of H. The rate of a reaction is written as

$$\log \underline{k}_X = \Psi \underline{v}_X + \underline{h}.$$

a. What is Ψ? What is $\underline{h}$?

b. The rates of acid-catalyzed ester hydrolysis reactions correlate very well with the above expression. What assumption of Taft does this result confirm?

c. The rates of base-catalyzed reactions also correlate well with the above expression, after due allowance for electronic effects. The Ψ_A (acid catalyzed) and Ψ_B (base catalyzed) values differ for a given substrate.

Reaction	Ψ_A	Ψ_B
$XCO_2C_2H_5 + 60\%\ C_2H_5OH/H_2O$	-1.12	-2.72
$XCO_2CH_2C_6H_5 + 60\%\ CH_3COCH_3/H_2O$	-2.36	-4.10

What assumption of Taft does this result call into question?

10. Related subject: introduction to kinetics

The preference of an electrophile for the meta or para position in attack of a given substrate is called the selectivity S and is expressed as

$$S = \log k_p/k_m ,$$

in which k_p/k_m is the ratio of the rates for respective attack at the para and the meta positions.

a. The selectivity for the bromination of toluene in acetic acid is much larger ($S = 2.64$) than that for the reaction of $(CH_3)_2CHBr/GaBr_3$ with toluene in benzene ($S = 0.55$). What is the cause of the large difference in S?

b. The bromination reaction rate in (a) correlates with σ^+ to give a ρ of -12.1, whereas the alkylation reaction correlates with σ to give a ρ of -2.3. Justify the σ correlations and the magnitudes of ρ in terms of mechanisms.

11. Related subjects: introduction to kinetics, orbital symmetry

a. 3-Substituted-2-methylisoquinolinium iodides (A) react with ethyl vinyl ether to give a Diels–Alder ([4 + 2]) adduct (B).

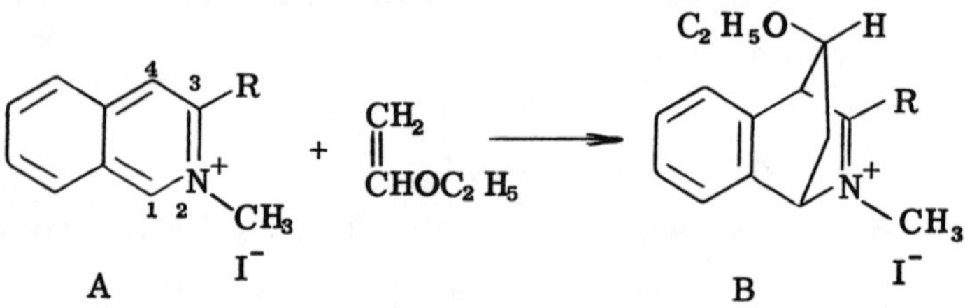

A Hammett plot (σ_m) of the rate of this reaction (R = NH_2, CH_3, H, C_6H_5, NHAc, Br) has a large, positive slope (ρ =

3.47, $\underline{r}$ = 0.979). Explain the substituent effect in terms of frontier molecular orbital (FMO) theory.

b. In the plot in part (a), the point for R = $\underline{tert}$-butyl is well off of the least-squares line for the other six points. There is a rate enhancement for this substituent of almost an order of magnitude. Why?

c. A similar acceleration is observed in the Diels-Alder reaction across the 6,11 position of certain acridizinium ions (C) with

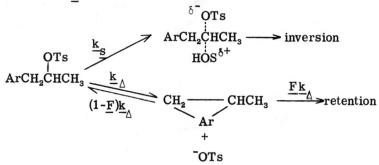

$^-ClO_4$ C

maleic anhydride. Thus at 65 °C, the presence of an 11-methyl group (R_1 = H, R_2 = CH_3) accelerates the reaction by a factor of 13.6 over the parent (R_1, R_2 = H). A 6-methyl group (R_1 = CH_3, R_2 = H), however, decelerates the reaction by a factor of 2. Explain these observations by reference to the activation parameters.

R_1	R_2	$\underline{E}_a$, kcal mol^{-1}	$\Delta \underline{S}^{\ddagger}$, eu
H	H	15.7	-33.1
H	CH_3	13.4	-34.7
CH_3	H	14.6	-37.6

12. a. 1-Aryl-2-propyl tosylates (A) can solvolyze by either solvent-assisted ($\underline{k}_\underline{s}$) or aryl-assisted ($\underline{k}_\underline{\Delta}$) pathways. The overall

titrimetric rate ($\underline{k}_t$) thus is given by

$$\underline{k}_t = \underline{k}_s + F\underline{k}_\Delta,$$

in which $\underline{F}$ is the proportion of aryl-assisted intermediate that goes on to product. A Hammett plot (below) of $\underline{k}_t$ for acetolysis is nonlinear.

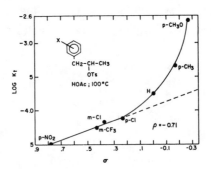

From the data in this plot, develop a method for determining the percent of aryl participation ($100\ \underline{Fk}_\Delta/\underline{k}_t$) and calculate these percentages.

b. The molecule $\underline{meso}$-1,4-diaryl-2,3-butyl ditosylate (B) can also solvolyze by $\underline{k}_s$ and $\underline{k}_\Delta$ pathways. Departure of the first

$$\text{B} \qquad \overset{\displaystyle \text{OTs}}{\underset{\displaystyle \text{OTs}}{\text{ArCH}_2\text{CH-CHCH}_2\text{Ar}}}$$

tosylate is rate determining. The nonlinearity of the Hammett plot (below) for B is much greater than for A. Using the

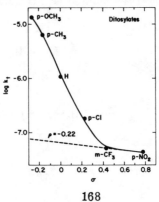

method developed in part (a), calculate the percentages of aryl participation for B and explain why the values are larger than those for A in most cases. An explanation for the larger percentages can be attempted even if no method for the calculation of the numbers was developed in (a).

• SOLUTIONS

1. a.

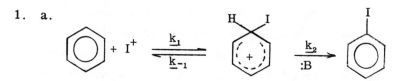

The large k_H/k_D suggests that the C–H bond is broken in the slow step, so that the second step is rate determining. The rate expression (see problems 6-1 and 6-2) then is $K_1 k_2$ times the concentration of the various reactants.

$$k_{obs} = K_1 k_2$$

$$\log k_{obs} = \log K_1 + \log k_2$$

$$\log \frac{k_{obs}}{k_{obs}^0} = \log \frac{K_1}{K_1^0} + \log \frac{k_2}{k_2^0}$$

$$\sigma \rho_{obs} = \sigma \rho_1 + \sigma \rho_2$$

$$\rho_{obs} = \rho_1 + \rho_2$$

The observed ρ is the sum of the ρ's for the individual steps since the rate and equilibrium constants are multiplicative. (The superscript zero in the above proof corresponds to the value when the substituent is H.) In the first step, an electron-withdrawing group should push the equilibrium to the left, so ρ_1 is negative. Since the substituent can conjugate directly with the positive charge, σ^+ is used, rather than σ. In the second step, positive charge decreases in the transition state, so an electron-withdrawing group should speed the reaction and ρ_2 is positive. The observed overall negative ρ indicates that ρ_1 must be dominant. A nonlinear Hammett plot is not expected in cases of this sort, since multiplicative constants lead to simple addition of ρ's. Had the first step been rate determining, the observed ρ would still have been negative, but probably larger. In multistep reactions like this, the composite nature of ρ means that it does not give a direct measure of electronic effects on the transition state in the slow step.

b. (i) For a maximal isotope effect, the force constants for C-H and H-B (base) stretching must be nearly equal, that is, the transition state $(C \cdots H \cdots B)$ is symmetrical. As X becomes more electron withdrawing (CO_2CH_3), the positive charge is less stable and the C-H bond is further broken in the transition state. As X becomes more electron donating (OCH_3), the positive charge is more stable and there should be less C-H bond breaking. In both directions from Cl, the transition state is less symmetrical, so k_H/k_D decreases. This explanation assumes that the second step is rate limiting and that secondary isotope effects on K_1 are negligible.

(ii) A change in the rate-determining step from the second to the first, i.e., from $k_2[B] \ll k_{-1}$ to $k_2[B] \gg k_{-1}$, (or a borderline situation) can also provide an explanation for the variation of k_H/k_D with X. Thus for Cl and CO_2CH_3, the second step is rate limiting, whereas for OCH_3, CH_3, and H, the first step is rate limiting. Normally, a change in the rate-determining step is accompanied by a nonlinear Hammett plot. A more detailed analysis of the rates and isotope effects led the authors to conclude that this explanation is not correct, and that the variation in k_H/k_D is due to changes in the symmetry of a single transition state (i).

Reference: L. L. Miller and B. F. Watkins, J. Am. Chem. Soc., 98, 1515 (1976).

2. a.

Ground state   Transition state

The correlation of any reaction rate with a solvent parameter such as ϵ, $\underline{Z}$, $\underline{E}_T$, or $\underline{Y}$ requires that the interactions of the solvent with the transition state closely resemble the interactions of the solvent with the defining system (capacitor for ϵ, solvolysis transition state for $\underline{Y}$, electronic excited state for $\underline{Z}$ or $\underline{E}_T$). Failure to exhibit a linear free energy correlation ($\log \underline{k}$) indicates that interactions on the microscopic scale (Kosower's cybotactic region) between the solvent and the transition state of the reaction under study differ in some way from those in the defining system. The values of $\underline{E}_T$ come from the long wavelength transition of the zwitterion A. The values of $\underline{E}$ come from the tetramethylguanidinium salt

A   B

of 2-cyano-5-nitrophenol (B), which is the product of the reaction. Solvent interactions with the transition state of the reaction must more closely resemble interactions with the excited state of this anion (B) than those with excited A. That solvent-anion interactions are the dominant consideration is supported by the correlation of the rates of 3-carboxybenzisoxazole decarboxylation with the rates of azide ion reaction with 4-nitrofluorobenzene, a nucleophilic aromatic substitution (correlation coefficient 0.985, slope 0.9).

b. The dipolar, aprotic solvents do not solvate anions well (see problems 8-6 and 8-7). The ground state is destabilized more than the transition state, since the negative charge in the latter case is more dispersed. The raising of the ground state energy and (in the present case) the possible lowering of the transition state energy through dispersion effects accelerates the rate of the reaction.

c. In the plot of log $\underline{k}$ vs. $\underline{E}$, the dipolar, aprotic solvents have the lowest $\underline{E}$ and the highest $\underline{k}$. In the middle of the plot come the ethers and hydrocarbons, i.e., materials of low dielectric constant. Finally, the hydroxylic solvents (traditionally of high polarity and high dielectric constant) show the lowest reactivity, with H_2O at the extreme. Certainly, solvent polarity has little direct influence on the reaction rate. The hydroxylic solvents can hydrogen bond to the ground state (C). Like the neutral carboxylic acid, this hydrogen-bonded species would be poorly reactive. Thus ground

C

state stabilization through hydrogen bonding slows the reaction. Since H_2O is more effective in this interaction, it exhibits the lowest reactivity.

d. The solvents H_2O and HMPA represent the extremes of transition state structure. The HMPA transition state has more negative charge than the hydrogen-bonded H_2O transition state, so the reaction in HMPA is more sensitive to substitution and ρ is more positive. The ρ's for most of the other solvents lie in the narrow range 1.75-2.10. Even the extremes are not vastly different. Thus the transition state structure probably changes very little with solvent. This conclusion is corroborated by the constancy of $\Delta \underline{S}^{\ddagger}$.

Reference: D. S. Kemp and K. G. Paul, J. Am. Chem. Soc., 97, 7305 (1975); D. S. Kemp, D. D. Cox, and K.G. Paul, ibid., 97, 7312 (1975).

3. a. The large negative ρ confirms that the transition state is electron deficient with respect to the ground state. Correlation with σ^+ implies that the positive charge can conjugate directly with the aryl ring in the transition state. Protonation must occur at the β carbon (the central allene carbon), since α or γ protonation would give products other than cinnamyl chloride. The structure of the high energy cationic inter-mediate should resemble that of the transition state, by the Hammond postulate. We will discuss intermediates, although strictly speaking, the data refer only to the transition state. Because the rate correlates with σ^+, charge development must occur at the α carbon. A cation of the type A is thereby eliminated. The cations A and B do not have allylic π overlap,

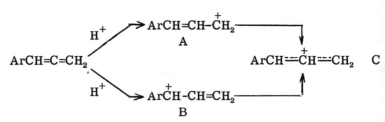

whereas C is planar and fully delocalized. If A can be eliminated, it remains to choose between B and C. Development of a full positive charge, as in the solvolysis of $ArC(CH_3)_2Cl$ or $ArCHCH_3Cl$, gives a ρ of about -4. The present allene gives a similar ρ. It would be expected that, to the extent that the cation is delocalized (as in C), the value of ρ should be diminished. Because no such diminution is observed, the first intermediate must be B, in which the allylic group is nonplanar and the positive charge is almost exclusively on the α position, as in B'. This structure most closely resembles the original allene, with the terminal

B'

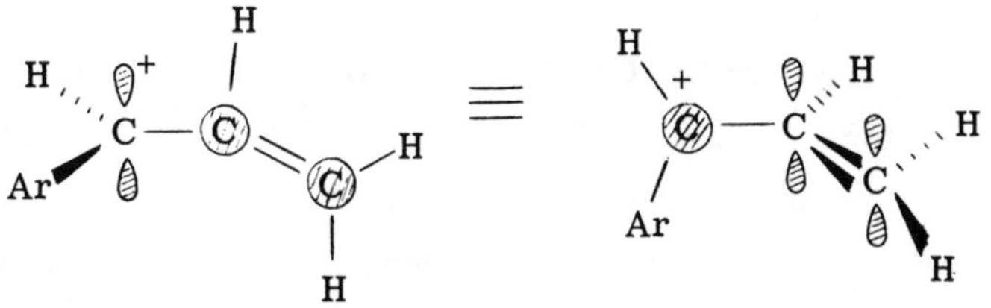

groups orthogonal. The first-formed structure B (B') then rotates rapidly to the planar structure C, from which products
are formed.

b. Because of the correlation with σ^+, the charge must be generated next to the ring for both systems. As in (a), the allene
aryl group is in conjugation with a full positive charge. The
first-formed intermediate again must be the orthogonal cation,
D in this case. The diminished value of ρ for the 1,3-buta-

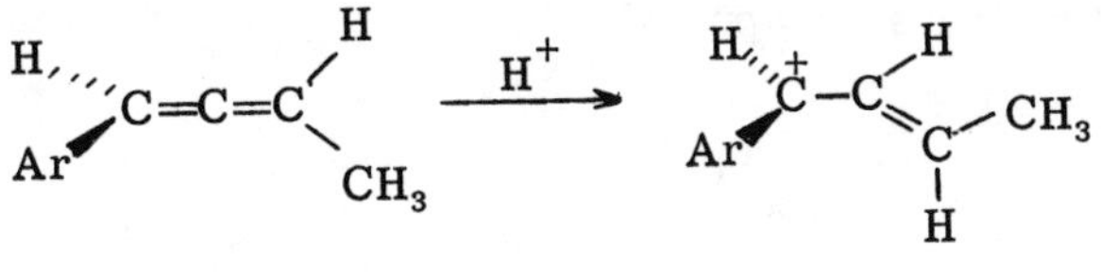

D

diene, however, indicates some delocalization of charge away
from the α position (E). Although charge is produced at the γ

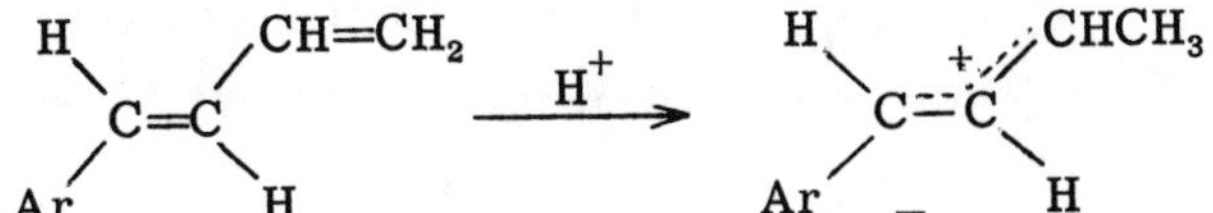

E

position, the ρ clearly indicates that it is also present at the
α position in the transition state. The 1,3-butadiene system
is already planar and hence requires no bond rotations to
achieve the delocalized allylic form E. The allene is initially
orthogonal and must have a 90° C—C bond rotation before the
planar form is reached. The present results indicate that this
rotation has not yet occurred in the transition state. Thus,
although the allene and the 1,3-butadiene give the same products, their initial transition states differ. The k_H/k_D for this
reaction is about 2, so that the initial protonation step is indeed

rate determining.

References: T. Okuyama, K. Izawa, and T. Fueno, J. Am. Chem. Soc., 95, 6749 (1973); K. Izawa, T. Okuyama, T. Sakagami, and T. Fueno, ibid., 95, 6752 (1973).

4. a. A concave downward Hammett plot is characteristic of a single mechanism with a change in the rate-determining step, rather than competition between two distinct mechanisms, which always gives a concave upwards plot. At least two steps are required for the present case, although more may be involved.

The most general rate law for the two step case, with the steady state approximation for [I], (see problems 6-1 and 6-2) is

$$\frac{d[B]}{dt} = \frac{k_1 k_2 [A]}{k_{-1} + k_2} .$$

When the second step is rate limiting ($k_2 \ll k_{-1}$), the right side becomes $K_1 k_2 [A]$. When the first step is rate limiting ($k_{-1} \ll k_2$), it is $k_1 [A]$. For electron-donating substituents ($R = CH_3O$, CH_3), the first step is rate determining, whereas for electron-withdrawing substituents ($R = Cl$, CF_3), the second step is rate determining.

b. One possible mechanism involves complexation of A with the Lewis acid, cyclization of one ring, and then cyclization of the other ring (three steps).

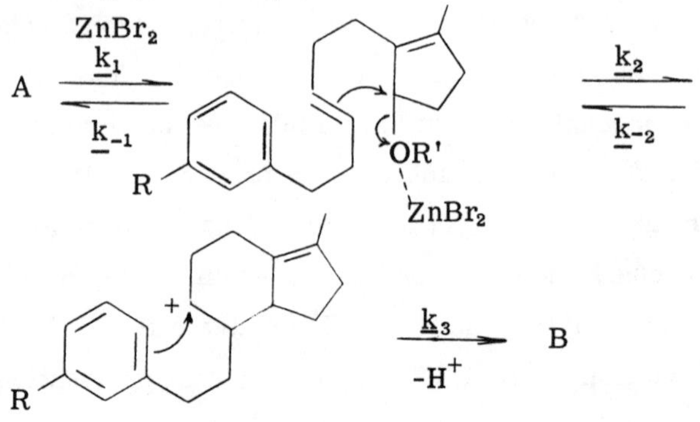

When $\sigma < 0$ (electron-donating groups), the last step, an electrophilic addition, is very rapid, so either the first (complexation) or second (cyclization of the ring that is further from the substituent R) step is rate determining. Because both the first and the second steps involve reactions that are far from the site of substitution, the rate is relatively insensitive to substitution, and ρ is close to zero. When $\sigma > 0$ (electron-withdrawing groups), the much slower electrophilic addition reaction becomes rate determining. A negative ρ is normal for a reaction in which positive charge moves closer to the aryl group in the transition state.

c. The question remains as to how many intermediate steps are necessary. The mechanism in (b) has two, complexation with the Lewis acid (k_1) and cyclization of the first ring (k_2). Let us first assume that both steps are necessary. The leaving group therefore has departed before the substituted aromatic ring participates in the cyclization of the second ring. It is observed, however, that the para-R/ortho-R ratio depends on the leaving group OR'. Thus the product-forming step must involve some sort of interaction between the aromatic ring and the leaving group. A three step mechanism with two distinct cyclizations has no such interaction. Therefore, it seems more likely that the extra step is unnecessary and that both rings form in a reasonably concerted fashion. All that is required is that there be some bond formation at both positions

in the transition state. The bonds need not be formed to the same extent.

Reference: P. A. Bartlett, J. I. Brauman, W.S. Johnson, and R. A. Volkmann, J. Am. Chem. Soc., 95, 7502 (1973).

5. a. The overall rate constant k_{obs} is the sum of the contributions from addition to either end of the double bond (k_1 for formation of ArĊHCHBrAr', k_2 for formation of ArCHBrĊHAr'). A given reaction experiences substituent effects from both aromatic rings. Thus formation of ArĊHCHBrAr' would have a ρ_1 from Ar (with σ_X^+) and a ρ_2 from Ar' (with σ_Y), and formation of ArCHBrĊHAr' would have a ρ_2 from Ar (with σ_X) and a ρ_1 from Ar' (with σ_Y^+). The ρ_1 is the reaction parameter for positive charge development on the carbon directly adjacent to the substituted ring (with σ^+), and the ρ_2 is the reaction parameter for positive charge development one carbon atom removed from the substituted ring (with σ).

$$k_{obs} = k_1 + k_2$$

$$\log\frac{k_1}{k_0} = \sigma_X^+ \rho_1 + \sigma_Y \rho_2$$

$$\log\frac{k_2}{k_0} = \sigma_X \rho_2 + \sigma_Y^+ \rho_2$$

$$\log\frac{k_{obs}}{k_0} = \log\frac{k_1 + k_2}{k_0 + k_0}$$

$$= \log\tfrac{1}{2}\left(\frac{k_1}{k_0} + \frac{k_2}{k_0}\right)$$

$$= \log\tfrac{1}{2}\left[10^{(\sigma_X^+ \rho_1 + \sigma_Y \rho_2)} + 10^{(\sigma_X \rho_2 + \sigma_Y^+ \rho_1)}\right]$$

This expression is clearly nonlinear. For cases in which this mechanism occurs, ρ_1 was found to be -5.1 and ρ_2 -1.4. The substituent effects are only approximately additive.

  b. When a symmetrical bromonium ion is formed, the Ar and Ar' rings must have identical effects (equal ρ's). Since the rings are not directly conjugated with the positive charge center, σ is used. The Hammett expression then is

$$\log \frac{k_{obs}}{k_0} = \sigma_X \rho + \sigma_Y \rho$$

$$= (\sigma_X + \sigma_Y)\rho.$$

This expression is linear.

c. The Hammett plot provides an operational method for distinguishing between the open carbonium ion (A) and the bromonium ion (B) mechanisms. In methanol, the curved Hammett plot indicates that either mechanism A or a combination of A and B occurs. Since methanol stabilizes carbonium ions well, and the rate data can be dissected into ρ_1 and ρ_2, the most likely mechanism is only A. In CCl_4, the linear Hammett plot indicates that only mechanism B occurs. This solvent is nonpolar, so the least polar, most delocalized transition state, i.e., that leading to the bromonium ion, is favored.

Reference: J. -E. Dubois and M. -F. Ruasse, <u>J. Org. Chem.</u>, <u>38</u>, 493 (1973).

6. a.

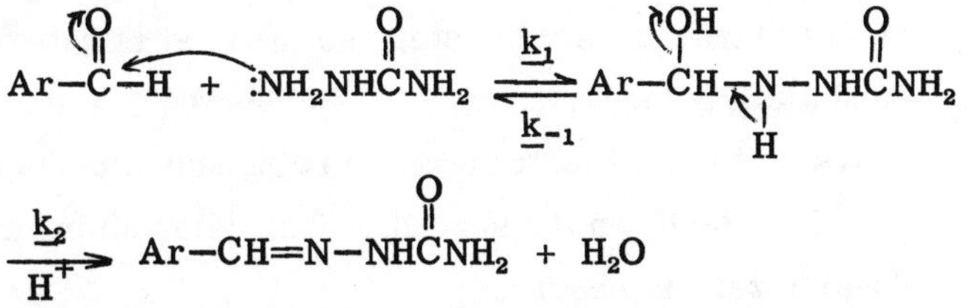

For simplicity, one step involving rapid proton exchange from nitrogen to oxygen was omitted from the mechanism. The nitrogen further away from the carbonyl group in semicarbazide is more nucleophilic, since its lone pair is not involved in amide resonance. As in any two step mechanism (see problem 9-4), the rate expression is either $\underline{K}_1\underline{k}_2$ [substrates] or $\underline{k}_1$ [substrates], depending on whether the second or the first step, respectively, is rate limiting.

b. The second step of the reaction is acid catalyzed. At high acidity $(\underline{k}_2[H^+] \gg \underline{k}_{-1})$ the second step is fast, so the first step is rate determining. The nucleophilic attack is assisted by electron withdrawal by the aromatic ring (making the site of attack more electrophilic), so ρ is positive.

c. At neutral pH, the acid-catalyzed second step becomes rate determining because of the reduced acidity $(\underline{k}_2[H^+] \ll \underline{k}_{-1})$. The rate law is $\underline{K}_1\underline{k}_2$ [substrates], so that the Hammett slope is $\rho_1 + \rho_2$, where ρ_1 is the reaction constant for the equilibrium $\underline{K}_1$ and ρ_2 is the reaction constant for the rate constant $\underline{k}_2$ (see problem 9-1). The equilibrium constant is increased by electron-withdrawing substituents (ρ_1 in fact is 1.81), since electron withdrawal disrupts aryl-carbonyl overlap on the left side but has little effect on the right. The rate constant of the second step is increased by electron donation (ρ_2 is -1.74), since conjugation is thereby increased in the transition state. Because these effects are equal and opposite, the net substituent effect is essentially nil.

d. At intermediate pH (3.9), the change from a rate-limiting first step to a rate-limiting second step is seen as a concave downward Hammett plot. Electron-donating substituents ($\sigma < 0$) favor the second step, so the first is rate determining and a fairly large, positive ρ is observed. A sharp break occurs at $\sigma = 0$. Electron-withdrawing substituents ($\sigma > 0$) favor the first step, so the second is rate determining and a ρ of nearly zero is observed.

Reference: B.M. Anderson and W. P. Jencks, J. Am. Chem. Soc., 82, 1773 (1960).

7. a. A Hammett plot with a concave downwards shape results from a change in the rate-determining step of a single mechanism. For electron-withdrawing groups ($\sigma > 0$, on the right), the carbonium ion intermediate (and the transition state leading to it) is destabilized, so that ionization (the second step) is rate determining. For electron-donating groups ($\sigma < 0$, on the left), this step is now quite fast. Ring closure of the stabilized carbonium is not favored, so that the cyclization (the third step) becomes rate determining. Thus $\underline{k}_{obs}$ is

either $\underline{K}_1\underline{k}_2$ (right side) or $\underline{K}_1\underline{K}_2\underline{k}_3$ (left side).

b. The ground state in the rate-determining third step for the
    substituents on the left has a direct interaction between the
    carbonium ion center and the substituent. Since the Hammett
    correlation is better with σ^+, this direct interaction must be
    maintained for the most part in the transition state. The
    question of why σ is used for the substituents on the right
    (second step rate determining) might be considered moot,
    since most of the substituents are electron withdrawing, for
    which $\sigma = \sigma^+$. Some, however, are electron donating, e.g.,
    CH_3 in $\underset{\sim}{8}$ and $\underset{\sim}{10}$, and both these points are on the line that uses
    σ. Compound $\underset{\sim}{10}$ occurs at the break, so it is present in the
    plot twice, with σ for the right and σ^+ for the left. The appli-
    cability of σ rather than σ^+ for the compounds on the right
    indicates that there is little direct conjugation between the
    substituents and the positive charge center. The transition
    state for these substrates must resemble the ground state of
    the rate-determining second step $(R\text{-}\overset{+}{O}H_2)$ rather than the
    product of this step, since the protonated alcohol offers no
    direct interaction between the substituent and the positive
    charge center.

c. On the left side (third step rate determining), positive charge
    is further removed from the aryl groups in the transition
    state than in the ground state of the third step, so ρ is positive.
    On the right side (second step rate determining), positive
    charge is closer to the aryl groups in the transition state to the
    second step, so ρ is negative.

d. Removal of a water molecule and rehybridization of the carbon
    from sp^3 to sp^2 in the second step should not have a large
    influence on $\Delta S^{\ddagger}$. For the compounds on the right, $\Delta S^{\ddagger}$ is
    found to lie in the small range from -1.0 to $+4.4$ eu. Cycliza-
    tion in the third step, however, requires proper alignment of
    the carbonium ion center and the biphenyl moiety. The loss
    of rotational degrees of freedom should give rise to a negative

$\Delta \underline{S}^{\ddagger}$. Thus the compounds on the left have the more negative $\Delta \underline{S}^{\ddagger}$ (-4.9 to -16.7 eu).

e. For compounds on the right (second step rate determining), the rate is dependent on the concentration of protonated alcohol. An increase in $[H^+]$ increases $[R\overset{+}{O}H_2]$ and hence the rate of the reaction. For compounds on the left (third step rate determining), the alcohol has been converted entirely to the carbonium ion prior to the rate-determining step, so there is little effect on the rate from $[H^+]$. The rate for compound 1 changes from 2.53×10^{-4} to 6.58×10^{-4} sec^{-1} when the concentration of H_2SO_4 goes from 4 to 6%, whereas the rate for compound 9 is 14.7×10^{-4} and 14.8×10^{-4} sec^{-1}, respectively, at the two concentrations.

f. Because of the equilibrium isotope effect (see problems 3-6 and 3-7), acids are generally about three times stronger in D_2O than in H_2O. The compounds on the left, whose rate is acid independent, therefore should show no change from H_2O to D_2O. Those on the right should react more rapidly in D_2O, since the equilibrium isotope effect raises the concentration of $R-\overset{+}{O}H_2$ $(R-\overset{+}{O}D_2)$. Thus the $\underline{k}_{D_2O}/\underline{k}_{H_2O}$ for compound 6 (on the right) is 3.10, whereas that for compound 9 (on the left) is 0.94.

Reference: H. Hart and E.A. Sedor, J. Am. Chem. Soc., 89, 2342 (1967).

8. a. The Taft equation includes one term for polar effects and

$$\log \frac{k}{\underline{k}_0} = \rho^* \sigma^* + \delta \underline{E}_{\underline{s}}$$

one term for steric effects. The substituent constant σ^* was developed to evaluate polar effects in aliphatic and ortho-substituted aromatic cases, with ester hydrolysis as the model reaction. The substituent constant $\underline{E}_{\underline{s}}$, determined from the rates of ester hydrolyses, is a measure of the steric effect of the substituent. The quantities ρ^* and δ are the respective sensitivities of a given reaction to polar and steric effects, compared to a value of unity for ester hydrolysis. The $\delta = 1.30$ for methanolysis of aliphatic menthyl esters indicates that the steric requirements of this reaction are only slightly larger than those of ester hydrolysis. The transition states (A and B) are very similar.

$$
\begin{array}{cc}
\overset{\delta^-}{} \overset{O}{\underset{\underset{\delta^- \ OCH_3}{\|}}{R-C-OC_{10}H_{16}}} & \overset{\delta^-}{} \overset{O}{\underset{\underset{\delta^- \ OH}{\|}}{R-C-OR'}}
\end{array}
$$

$$
\qquad A \qquad\qquad\qquad B
$$

The facts that $^-OCH_3$ is slightly larger than ^-OH and that menthyl is larger than the defining R' groups probably contribute to the small increase.

b. The identity of the Hammett ρ in the aryl series with the Taft ρ^* in the aliphatic series indicates that the polar and resonance effects are about the same for the two types of substrates. This result serves to corroborate Taft's assumption that the electronic effects for aliphatic substrates can be separated from the steric effects by subtracting the $\log \dfrac{k}{k_0}$ for the acid-catalyzed reaction from that for the base-catalyzed reaction.

References: C. G. Mitton, R. L. Schowen, M. Gresser, and J. Shapley, J. Am. Chem. Soc., 91, 2036 (1969); J. Shorter, Quart. Rev. (London), 24, 433 (1970).

9. a. The slope Ψ is a measure of the sensitivity of the reaction to the function on the ordinate, in this case the steric substituent parameter v_X. All linear free energy relationships of this form yield a sensitivity factor from the slope (ρ, m, δ). The intercept h is the value of $\log k_X$ when the substituent is H. Since $v_H = 1.20-1.20 = 0$ for H, the coefficient of Ψ is zero, and $\log k_H = h$. This term is necessary because Charton used a $\log k_X$ expression, whereas Hammett and Taft used a $\log k_X/k_H$ expression.

b. The parameter $\underline{v}_X$, derived from the van der Waals radius, depends only on steric properties. Taft assumed that the acid-catalyzed ester hydrolysis reaction depends only on steric factors, since $\rho \sim 0$. The successful correlation of these rates with $\underline{v}_X$, without allowance for polar or resonance (including hyperconjugative) effects, substantiates the Taft assumption.

c. In order to evaluate the polar/resonance effects in aliphatic systems, Taft subtracted the log $\underline{k}/\underline{k}_0$ for the acid-catalyzed ester hydrolysis from that for the base-catalyzed reaction (see problem 9-8). If the steric effect is the same in the two reactions and if the acid-catalyzed reaction has no polar/resonance influence (see part (a)), then this subtraction provides a number (eventually σ^*) that is a measure of the polar/resonance effects in the base-catalyzed systems. Charton measured Ψ for both acid- and base-catalyzed ester hydrolysis and found that they are statistically different in many cases. The two sets of data given in part (c) of the question are among the worst. Better agreement is sometimes observed. Thus for $XCO_2CH_2CH_3 + 70\%\ CH_3COCH_3/H_2O$, $\Psi_A = -2.49$ and $\Psi_B = -2.65$. Charton concludes that Taft's assumption of a constant steric effect in acid and in base is at best approximate and is at times false. The favorable result in problem 9-8 is not in contradiction with these results, since the assumption is sometimes valid. The acid- (A) and base-catalyzed (B) transition states differ on two counts. There are two more protons

A   B

in A (as Taft pointed out) and A has a positive charge, B a negative charge. The values of Ψ_A and Ψ_B suggest that the base-catalyzed reaction has the greater steric effect, so the number of protons cannot be significant. The differences in charge type could lead to solvation differences that are reflected in the transition state as steric perturbations. Alternatively, the acid and base transition states could occur at different points on the reaction coordinate. If the transition state of the base-catalyzed reaction is closer to the crowded tetrahedral intermediate than that of the acid-catalyzed case, a larger steric effect would be expected. A decision from among these explanations has not yet been reached.

References: M. Charton, J. Am. Chem. Soc., 97, 1552, 3691 (1975).

10. a. The differences in S result from differences in the stabilities of the respective electrophiles. For the bromination of toluene in acetic acid, the electrophile (probably Br^+) is relatively stable and hence is highly selective (larger S). Because the CH_3 substituent in toluene is ortho, para directing, more of the preferred position of substitution (para rather than meta) is observed when the electrophile is more selective. The electrophile for the alkylation of toluene in benzene is $CH_3\overset{+}{C}HCH_3$ (probably complexed in some way to $^-GaBr_4$). This species is relatively unstable in the nonpolar solvent benzene, so it is highly reactive and poorly selective (lower S). The natural ortho, para direction is lessened to some extent when the electrophile is very reactive.

b. The electrophilic substitution mechanism leading to a σ complex involves a transition state (similar to the σ intermediate)

that has considerable positive charge on the aromatic ring. Because substituents attached to the ring at appropriate positions could conjugate directly with the positive charge, a σ^+ correlation should be observed. Because of the large positive charge density in the transition state, ρ should be large and negative. The observed σ^+ correlation with a very large, negative ρ in the bromination reaction is thus taken to be indicative of a σ complex. If the transition state resembles the starting material or a π complex (A), there is little positive

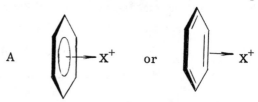

charge on the aromatic ring. Consequently, a correlation with σ and a negative ρ of modest magnitude should be observed. The characteristics of the alkylation reaction (σ correlation, small negative ρ) are therefore in accord either with a π complex mechanism or a very exothermic reaction (transition state resembling starting material by the Hammond postulate). The π complex mechanism appears to be the more likely possibility. Any electrophilic substitution reaction can involve both σ and π intermediates consecutively (starting material $\rightleftharpoons$ π complex $\rightleftharpoons$ σ complex $\rightleftharpoons$ product). The response to σ vs. σ^+ and the magnitude of ρ can therefore assist in ascertaining the identity of the highest transition state.

References: L. M. Stock and H.C. Brown, J. Am. Chem. Soc., 81, 3323 (1959); Advan. Phys. Org. Chem., 1, 35 (1963); for an alternative view of these reactions, see J.H. Ridd, Acc. Chem. Res., 4, 248 (1971).

11. a. When the dienophile has an electron-donating substituent, its HOMO and LUMO are relatively high, as in the diagram. Consequently, the important interaction is between the HOMO of the dienophile and the LUMO of the diene. The positive slope of the Hammett plot indicates that the reaction is accel-

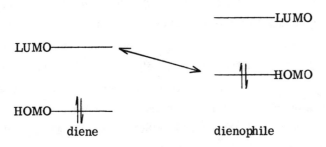

erated by electron-withdrawing groups on the diene. Electron-
withdrawing groups lower the HOMO and LUMO energies for
the diene, thereby intensifying the HOMO (diene)-LUMO (dienophile) interaction. For dienes with rather higher HOMO/
LUMO energies or for dienophiles with electron-withdrawing
groups and hence rather lower HOMO/LUMO energies, the
principal interaction is between the HOMO of the dienophile
and the LUMO of the diene.

b. A steric effect of the tert-butyl group must overcome the FMO
electronic effects. The acceleration most likely is caused by
relief of peri strain experienced by the proton at position 4.
The interaction of this proton with the peri proton and the
tert-butyl group (buttressed by the 2-methyl group) is decreased as the ring becomes nonplanar in the transition state.
The authors confirmed this steric interaction in the acridizinium system (see part (c)) by observing that the 7,10,11-
trimethyl system reacts over 100 times faster than the unsubstituted system. In this case, the primary peri interaction
is between the two methyl groups at the 10 and 11 positions.

c. A single methyl group at the 11 position loses the peri interactions with the 1 and 10 protons on reaction with the dienophile. This relief of steric strain gives rise to the observed
rate acceleration of 13.6. The strain relief is accompanied by
a decrease of 2.3 kcal mol^{-1} in the Arrhenius activation energy and little or no change in the activation entropy. The anomalous rate deceleration for the 6-methyl derivative is also
accompanied by an enthalpic decrease, but of only 1.1 kcal

mol^{-1}. An entropic decrease of 3.5 eu, however, is more than enough to offset the favorable enthalpy change and cause a rate deceleration. The authors are not explicit about the nature of the entropy effect, only saying that "the molecular vibrations of [the methyl group] reduce the number of pathways by which a maleic anhydride molecule can make a fruitful approach." Why the 6-methyl but not the 11-methyl shows the effect is not discussed. The larger enthalpic effect of the 11-methyl group suggests that bonding is more developed to the 11 than to the 6 position.

References: C. K. Bradsher, T. G. Wallis, I. J. Westerman, and N. A. Porter, J. Am. Chem. Soc., 99, 2588 (1977); I. Fleming, "Frontier Orbitals and Organic Chemical Reactions," Wiley-Interscience, London, 1976, pp 110-13.

12. a. The fraction of aryl participation is given by

$$\frac{F k_{\Delta}}{k_t} = 1 - \frac{k_s}{k_t}$$

$$= \frac{k_t / k_s - 1}{k_t / k_s} \ .$$

Thus knowlege of k_t / k_s will produce the desired ratio directly. Strongly electron-withdrawing groups will completely eliminate the k_{Δ} pathway, whose transition state possesses considerable build-up of positive charge. These substituents, p-NO$_2$, m-CF$_3$, m-Cl, and p-Cl, in fact give a reasonably linear Hammett plot. The line from these specific groups must correspond to the k_s process alone. Extrapolation (dotted line in the plot) can give a calculated k_s for all the other substituents. The difference between the solid and the dotted lines in the plot therefore is $\log(k_t / k_s)$, from which $F k_{\Delta} / k_t$ can be calculated according to the above equation.

X	p-CH$_3$O	p-CH$_3$	H	p-Cl	m-Cl	m-CF$_3$	p-NO$_2$
100 $\underline{F}\underline{k}_\Delta/\underline{k}_t$	91	67	36	0	0	0	0

For 1-aryl-2-propyl tosylates under conditions of solvolysis, aryl participation becomes significant at H and increases with the more electron-donating p-CH$_3$ and p-OCH$_3$.

b. The percentages calculated from the plot are given below.

X	p-OCH$_3$	p-CH$_3$	H	p-Cl	m-CF$_3$	p-NO$_2$
100 $\underline{F}\underline{k}_\Delta/\underline{k}_t$	99	99	94	68	0	0

Clearly, the second tosyloxy group causes a greater amount of aryl participation. During the rate-determining loss of the first tosyloxy group, the second tosyloxy group exerts an inductive/field influence. Polar withdrawal of electrons by the second toxyloxy group destabilizes the development of positive charge at the site of removal of the first tosyloxy group. The greater electron demand at this position encourages stronger participation by the aryl group. Electron delocalization via aryl participation relieves the polar destabilization of the transition state caused by the remaining tosyloxy group. This phenomenon has been termed "inductive enhancement of solvolytic participation."

References: C. J. Lancelot and P. v. R. Schleyer, J. Am. Chem. Soc., 91, 4291 (1969); J. B. Lambert, H. W. Mark, and E. S. Magyar, ibid., 99, 3059 (1977); F. L. Schadt III, C. J. Lancelot, and P. v. R. Schleyer, ibid., 100, 228 (1978). The 1978 paper refines the 100 $\underline{F}\underline{k}_\Delta/\underline{k}_t$ figures. The 1969 figures have been retained here, however, in order to have a common basis for calculation of the figures for systems A (1969) and B (1977).

ALSO SEE PROBLEMS 10-1, 10-4, 10-5, 10-6, 10-10.

10
HOMOGENEOUS CATALYSIS

PROBLEMS

1. Related subject: kinetic electronic and steric effects

   Consider the nucleophilic displacement of chloride in an aryl sulfonyl chloride by an aniline nitrogen (methanol solvent). Either ring may have substituents. The second aniline molecule in the equation simply serves as a base.

   $$ArSO_2Cl + 2H_2\overset{..}{N}Ar' \longrightarrow ArSO_2\overset{..}{N}HAr' + H_3\overset{+}{N}Ar'Cl^-$$

   a. Hammett plots may be constructed by varying the substituents in either ring (Ar or Ar', but not both). What should be the approximate sign and magnitude of the two ρ's?

   b. A Brønsted plot may be constructed by varying the substituent on Ar' (with Ar constant) and plotting $\log \underline{k}$ vs. $\underline{pK}_a$ (for the conjugate acid of the amine). If such a plot is made for each member of the Ar series, different values of β are obtained, e.g., for p-NO$_2$ (in Ar) β is 0.93, for p-OCH$_3$ β is 0.65. What is the significance of this variation?

2. <u>Related subject: introduction to kinetics</u>

   a. The base-catalyzed decomposition of nitramide (NH$_2$NO$_2$) involves removal of a proton from the substrate. There is one Brønsted plot for neutral bases, a parallel plot transposed to faster rate for positively charged bases, and a third parallel line transposed to lower rates for negatively charged bases (charge of unity on the bases). Explain.

   b. The hydrolysis of ethyl vinyl ether involves transfer of a proton to the substrate. Positively charged acids (amino acid cations) give a Brønsted plot parallel to but about two orders of magnitude below that of neutral (carboxylic) acids, and negatively charged acids (HSO$_4^-$, HCO$_2$CO$_2^-$) are above the line for the neutral acids. Explain.

   c. Predict what might be the result when a Brønsted plot is constructed from a series of neutral acids, some of which have highly polar substituents.

3. <u>Related subjects: introduction to kinetics, kinetic isotope effects</u>

   a. Consider the enolization of acetone under basic, aqueous conditions. If an acidic material such as acetic acid is present, one possible mechanism is as follows.

$$CH_3\overset{\overset{O}{\|}}{C}CH_3 + {}^-OH \underset{}{\overset{K_1}{\rightleftharpoons}} {}^-CH_2\overset{\overset{O}{\|}}{C}CH_3 + H_2O$$

$$CH_2{=}\overset{\overset{O^-}{|}}{C}CH_3 + HOAc \xrightarrow{k_2} CH_2{=}\overset{\overset{OH}{|}}{C}CH_3 + {}^-OAc$$

Is the reaction as written subject to general or specific base catalysis? Justify your answer by an appropriate kinetic derivation.

   b. The Dawson-Spivey third order term ($k_{HOAc}/{}^-OAc$[HOAc]·[${}^-$OAc][acetone]) in the bromination or enolization of acetone (see problem 6-2) has the following properties.

     (i) The k_H/k_D (CH$_3$COCH$_3$ vs. CD$_3$COCD$_3$) is 5.8.

(ii)   The solvent k_{H_2O}/k_{D_2O} is 2.0 for both CH_3COCH_3 and CD_3COCD_3.

(iii)   The Brønsted β_{AB} is 0.15, i.e., log $\underline{k}$ vs. log $\underline{K}_a$ is almost horizontal for a series of HA/A^- buffers.

Describe the mechanism of the third order reaction, utilizing each piece of data.

4. <u>Related subjects</u>: <u>introduction to kinetics, kinetic isotope effects, kinetic electronic and steric effects</u>

The addition of water to $ArC\equiv CNR_2$ (ynamines) in basic medium is general acid catalyzed (Brønsted $\alpha = 0.74$). The solvent isotope effect k_{H_2O}/k_{D_2O} is 4, the Taft ρ^* (varying R on nitrogen) is -2.2, the $\Delta\underline{S}^\ddagger$ is -39 eu. What is the mechanism? Discuss the applicability of each piece of information.

5. <u>Related subjects</u>: <u>kinetic isotope effects, kinetic electronic and steric effects</u>

Two mechanisms have been considered for the acid-catalyzed hydrolysis of phosphinamides.

$$R_2\overset{\overset{O}{\|}}{P}-NR_2' + H_2O \xrightarrow{H^+} R_2\overset{\overset{O}{\|}}{P}-OH + HNR_2'$$

a. <u>O</u>-Protonation/pentacoordinate intermediate

$$R_2\overset{\overset{O}{\|}}{P}-NR_2' \xrightarrow{H^+} R_2\overset{\overset{OH}{\|}}{\underset{+}{P}}-NR_2' \xrightarrow{H_2O}$$

$$R_2\overset{\overset{OH}{\|}}{\underset{\overset{\|}{+OH_2}}{P}}-NR_2' \longrightarrow R_2\overset{\overset{OH}{\|}}{\underset{\overset{\|}{OH}}{P}}\overset{+}{-NHR_2'} \xrightarrow{-NHR_2'}$$

$$R_2\overset{\overset{OH}{\|}}{\underset{+}{P}}-OH \xrightarrow{-H^+} R_2\overset{\overset{O}{\|}}{P}-OH$$

b. <u>N</u>-Protonation/direct displacement

$$R_2\overset{\overset{O}{\|}}{P}-NR_2' \xrightarrow{H^+} R_2\overset{\overset{O}{\|}}{P}-\overset{+}{N}HR_2' \xrightarrow[-NHR_2']{H_2O}$$

$$R_2\overset{\overset{O}{\|}}{P}-\overset{+}{O}H_2 \xrightarrow{-H^+} R_2\overset{\overset{O}{\|}}{P}-OH$$

The reaction is subject to specific acid catalysis, the solvent isotope effect k_{H_2O}/k_{D_2O} is 0.80, and the Taft ρ^* for variation of the R' group is -1.0. The rate of an open chain compound $(R = \underline{i}\text{-}C_3H_7\text{-})$ is 10^3 faster than that of a small ring substrate $(R, R = \text{-}CH_2CH_2CH_2\text{-})$. On the basis of this evidence which mechanism is probably correct, and which step of the correct mechanism is rate limiting? State clearly which data are useful in differentiating between mechanisms and which data are useful in designating the rate-determining step.

6. Related subjects: kinetic isotope effects, kinetic electronic and steric effects

   a. The base-catalyzed iodination of the acetyl methyl group in acetylpyridines A-D $(Ac = CH_3CO)$ is zero order in $[I_2]$ and

          A   B   C   D

subject to general base catalysis, and the kinetic isotope effect $\underline{k}_H/\underline{k}_D$ $(CH_3$ vs. $CD_3)$ is about 7. Briefly outline the mechanistic significance of these results.

   b. The following rates were observed in acetate buffer $(\underline{k}$ in $\ell\ \text{mol}^{-1}\text{sec}^{-1})$. Why is A accelerated so much more than B by $\underline{N}$ methylation?

A	B	C	D
0.73×10^{-5}	3.4×10^{-5}	830×10^{-5}	46×10^{-5}

   c. Complexation of A with Zn^{2+} leads to a 4.5×10^3 acceleration, with Cu^{2+} 2×10^5. Complexation of B does not affect the rate. Complexed A still is general base catalyzed and exhibits a large primary isotope effect. What is the role of metal catalysis in the mechanism?

7. Related subject: kinetic isotope effects

Water adds across the N3-C4 double bond of pteridine in a reaction that is catalyzed by both acid and base.

a. At low pH, the reaction is specific acid catalyzed, and k_{H_2O}/k_{D_2O} is 0.5. Suggest a mechanism.

b. At neutral pH, k_{H_2O}/k_{D_2O} is 3.4. The catalytic coefficient for H_2O is nearly 10^3 faster than that predicted by the Brønsted plot. Suggest a mechanism.

c. At high pH, the reaction is general base catalyzed, and k_{H_2O}/k_{D_2O} is 0.7. Suggest a mechanism.

8. Related subjects: <u>introduction to kinetics, kinetic solvent effects</u>
Hydration of alkenes is normally first order in both alkene and acid. Two mechanisms are possible, the A-1, in which the proton rapidly forms a π complex before becoming attached to carbon, and the A-S_E2, in which the proton is transferred directly to carbon in the slow step.

a. Which mechanism requires general acid catalysis and why?

b. Hydration of 2,3-dimethyl-2-butene and of <u>trans</u>-cyclooctene gave the dependencies on the $H_3PO_4/H_2PO_4^-$ buffer concentration presented in the plots (p. 196). Which mechanism (A-1 or A-S_E2) do the data favor and why?

c. What is the purpose of the correction for constant $[H^+]$ in the plots?

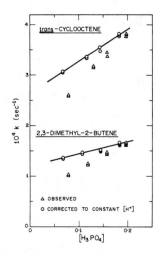

9. <u>Related subject: ionic equilibria</u>

Consider the acid-catalyzed hydrolysis of lactones.

$$(H_2C)_n \overset{O}{\underset{O}{\big\backslash}} C \quad + \quad H_2O \quad \rightleftharpoons \quad HO\text{-}(CH_2)_n\text{-}\overset{O}{\underset{||}{C}}\text{-}OH$$

The rate expression for β-propiolactone (A) is

rate = $\underline{k}$[lactone]$\underline{h}_0$.

The expression for γ-butyrolactone (B) is

rate = $\underline{k}$[lactone][H^+].

A   B

a. With the use of the Zucker-Hammett hypothesis, suggest a plausible mechanism for the hydrolysis of β-propiolactone. Label the rate-determining step.

b. Do the same with γ-butyrolactone.

c. The acid-catalyzed formation of γ-butyrolactone from γ-hydroxybutyric acid, $HOCH_2CH_2CH_2CO_2H$, should follow the microscopic reverse of your mechanism in (b). Should the rate of lactonization be proportional to $\underline{h}_0$ or [H^+]?

10. Related subjects: ionic equilibria, kinetic isotope effects, kinetic electronic and steric effects

Enol acetates may hydrolyze in aqueous H_2SO_4 either by the normal (tetrahedral intermediate) ester mechanism or by a proton transfer to the double bond ($A-S_E2$) (also see problem 10-8).

$$\text{Normal:} \quad CH_2{=}CAr{-}O\overset{O}{\overset{\|}{C}}CH_3 + H^+ \rightleftharpoons CH_2{=}CAr{-}O\overset{+OH}{\overset{\|}{C}}CH_3$$

$$\xrightarrow[H_2O]{\text{slow}} CH_2{=}CAr{-}O\overset{OH}{\underset{|}{C}}{-}CH_3 \xrightarrow{\text{fast}} CH_2{=}CAr{-}^+\overset{H}{\overset{|}{O}}{-}C(OH)_2CH_3$$

$$\xrightarrow[-H^+]{\text{fast}} CH_2{=}CAr{-}\overset{+OH_2}{\underset{|}{OH}} + CH_3CO_2H \xrightarrow{\text{fast}} CH_3{-}\overset{O}{\overset{\|}{C}}{-}Ar + CH_3CO_2H$$

$$A-S_E2: \quad CH_2{=}CAr{-}O\overset{O}{\overset{\|}{C}}CH_3 + H^+ \xrightarrow{\text{slow}} CH_3{-}\overset{+}{C}Ar{-}O\overset{O}{\overset{\|}{C}}CH_3$$

$$\xrightarrow[H_2O]{\text{fast}} CH_3{-}\overset{+OH_2}{\underset{|}{C}}Ar{-}O\overset{O}{\overset{\|}{C}}CH_3 \xrightarrow{\text{fast}} CH_3{-}\overset{OH}{\underset{|}{C}}Ar{-}^+\overset{H}{\underset{|}{O}}{-}\overset{O}{\overset{\|}{C}}{-}CH_3$$

$$\xrightarrow{\text{fast}} CH_3{-}^+\overset{OH}{\underset{|}{C}}{-}Ar + CH_3CO_2H \xrightarrow[-H^+]{\text{fast}} CH_3{-}\overset{O}{\overset{\|}{C}}{-}Ar + CH_3CO_2H$$

Variation of the substituent on the Ar group gives a curved Hammett plot (concave upwards), with $\rho = -1.9$ for $\sigma^+ < 0$ and $\rho = 0.0$ for $\sigma^+ > 0$ (6% H_2SO_4, $H_0 = 0$). The rates of the individual substrates vary with the acidity. For Ar = p-$CH_3OC_6H_5$, log k vs. H_0 is linear and k_{H_2O}/k_{D_2O} is 2.50. For Ar = p-$NO_2C_6H_5$, log k vs. H_0 is curved, with $k_{H_2O}/k_{D_2O} = 0.75$ at low acidities ($H_0 = 0$ to -2) and 3.25 at high acidities ($H_0 = -4$ to -6). Rationalize these data (Hammett plot, H_0 dependence, isotope effect) in terms of the mechanisms. Interpret the differences between p-CH_3O and p-NO_2 mechanistically, and explain why these differences occur.

11. Related subject: introduction to kinetics

Pyridine nucleophiles react with 2,4-dinitrophenyl methyl carbonate to displace the phenolate ion. The observed rate, adjusted

$$XH_4C_5N: \ + \quad \underset{CH_3O}{\overset{O}{\underset{\|}{C}}}\ OC_6H_3(NO_2)_2 \quad \underset{\underline{k}_{-1}}{\overset{\underline{k}_1}{\rightleftharpoons}}$$

$$XH_4C_5\overset{+}{N}-\underset{OCH_3}{\overset{O^-}{\underset{|}{\overset{\|}{C}}}}-OC_6H_3(NO_2)_2 \overset{\underline{k}_2}{\longrightarrow} \quad \underset{XH_4C_5\overset{+}{N}}{\overset{O}{\overset{\|}{C}}}\ OCH_3 \ + \ {}^-OC_6H_3(NO_2$$

for reaction with solvent, is first order in substrate and nucleophile

$$\frac{d[\text{phenolate}]}{dt} = \underline{k}_{obs}\,[N][S]$$

$$= \frac{\underline{k}_1\underline{k}_2[N][S]}{\underline{k}_{-1} + \underline{k}_2}$$

The Brønsted plot of $\underline{k}_{obs}$ vs. the $\underline{pK}_a$ for the conjugate acid of the substituted pyridine is curved. Provide an explanation for the

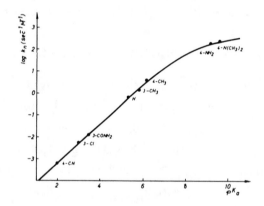

curvature in terms of the above mechanism. Include in your discussion the significance of the point of maximum curvature ($\underline{pK}_a = 7.8$). The $\underline{pK}_a$ of 2,4-dinitrophenol is about 4.0.

12. Related subjects: ionic equilibria, kinetic solvent effects

Acid-catalyzed hydration of alkenes via rate-determining proton transfer $(A-S_E2)$ to give trigonal carbocations occurs at rates that are similar to or slower than those of hydration of analogous alkynes to give digonal (vinyl) carbocations. These results are

$$\begin{array}{c}\diagdown\\ \diagup\end{array}C=C\begin{array}{c}\diagup\\ \diagdown\end{array} + H_3O^+ \longrightarrow \ H-\overset{|}{\underset{|}{C}}-\overset{+}{C}\diagup + H_2O \longrightarrow \text{products}$$

$$-C\equiv C- \ + H_3O^+ \longrightarrow \ \begin{array}{c}\diagdown\\ H\diagup\end{array}C=\overset{+}{C}- \ + H_2O \longrightarrow \text{products}$$

surprising, since the trigonal cations are 10-20 kcal mol^{-1} stabler than the digonal cations and since other electrophilic additions, e.g., halogenation, are indeed much faster for the alkenes. One explanation has been that digonal cations are much better stabilized by solvation in the transition state than are trigonal cations. A transition state solvation parameter $\Phi_{\ddagger}$ can be defined in an analogous fashion to the equilibrium solvation factor Φ (see problems 5-4 and 5-5). For the A-S_E2 mechanism, the definition takes the form

$$\log \underline{k} + \underline{H}_0 = \Phi_{\ddagger}(\underline{H}_0 + \log [H^+]) + \log \underline{k}_0 .$$

From plots of $(\log \underline{k} + \underline{H}_0)$ vs. $(\underline{H}_0 + \log [H^+])$, the following data were obtained.

	$\Phi_{\ddagger}$(alkyne)	$\Phi_{\ddagger}$(alkene)
p-CH$_3$OC$_6$H$_4$C≡CH	-0.67	
p-CH$_3$OC$_6$H$_4$CH=CH$_2$		-0.54
p-CH$_3$C$_6$H$_4$C≡CH	-0.58	
p-CH$_3$C$_6$H$_4$CH=CH$_2$		-0.34
C$_6$H$_5$C≡CH	-0.44	
C$_6$H$_5$CH=CH$_2$		-0.31
p-ClC$_6$H$_4$C≡CH	-0.29	
p-ClC$_6$H$_4$CH=CH$_2$		-0.28
C$_2$H$_5$C≡CC$_2$H$_5$	-0.25	
trans-C$_2$H$_5$CH=CHC$_2$H$_5$		-0.34

What are the mechanistic implications of these results? Comment on the change in $\Phi_{\ddagger}$ as the aryl substituent varies within a series.

SOLUTIONS

1. a. In the transition state (A), positive charge is building up on

A

the nitrogen atom adjacent to the Ar' ring. Electron-donating Ar' substituents (Y) therefore accelerate the reaction, so that the ρ measured for each X in Ar by varying Y in Ar' should be negative and reasonably large. The observed values of ρ were about -2. On the other hand electron-withdrawing groups (X) on the Ar ring make the sulfonyl site more electron deficient and hence more susceptible to nucleophilic attack. The ρ measured for each Y by varying X should be positive but probably not very large. The observed values were about +1.

   b. In this context, β provides a measure of the extent of bonding between the nucleophile and the sulfonyl chloride, just as α is a measure of S-H bond making in a proton transfer reaction (S + H-A). The linearity of the Brønsted plot illustrates the relation between basicity (ability of the nitrogen lone pairs to bond <u>completely</u> to a proton, as measured by the pK_a of the conjugate acid) and nucleophilicity (ability of the lone pair to bond <u>partially</u> to sulfur in the transition state to nucleophilic displacement, as measured by log $\underline{k}$). The lower the acidity (higher the pK_a), the stronger the base and hence the nucleophile. The variation of β with the substituent X on the arene-sulfonyl ring (Ar) is the result of different stages of bonding in the transition state. The β of 0.93 for X = $\underline{p}$-NO$_2$, compared to 0.65 for $\underline{p}$-OCH$_3$, indicates that S---N bond formation is greater in the former case. Electron withdrawal causes sulfur to be more electron deficient and more in need of bond formation to the incoming nucleophile. This is the normal behavior for an S_N2 transition state. A plot of the Brønsted β vs. σ (for each X substituent, β determined by variation of Y) is linear, as is a plot of the Hammett ρ vs. log $\underline{K}_a$ (for each Y substituent, ρ determined by variation of X). The slope of both plots is 0.26. The ρ's and β's are interrelated in a definable fashion.

Reference: O. Rogne, <u>J</u>. <u>Chem</u>. <u>Soc</u>. <u>B</u>, 1855 (1971).

2. a.

Neutral base	$\overset{\delta^+}{B:}----H----\overset{\delta^-}{NH}-NO_2$
Positively charged base	$^+B:----H----\overset{\delta^-}{NH}-NO_2$
Negatively charged base	$^-B:----H----\overset{\delta^-}{NH}-NO_2$

Since the lines in the plots are parallel, β is the same for all three base charge types and the extent of proton transfer is similar. In each case, the substrate (nitramide) develops a partial negative charge in the transition state, because of removal of the proton. Approach of a positively charged base is therefore more favorable (faster) than that of a neutral base with the same $p\underline{K}_b$. Similarly, approach of a negatively charged base is less favorable (slower).

b. The same phenomenon, except in reverse, occurs here. The substrate gains a partial positive charge in the transition state by protonation of carbon, as illustrated for a neutral general acid, HA. Approach of a positively charged acid

$$\overset{\delta^-}{A}----H----\overset{\delta^+}{CH_2}====CH====\overset{\delta^+}{O}-CH_2CH_3$$

is less favorable (slower) than that of a neutral acid with the same $p\underline{K}_a$. Approach of a negatively charged acid is more favorable (faster).

c. The examples of (a) and (b) show that electrostatic phenomena can exert significant influences on Brønsted plots. Therefore it is likely that strongly dipolar substituents might have similar effects, although monopole-dipole interactions would

be smaller than the monopole-monopole interactions of parts
(a) and (b). The result would be that highly polar acids (or
bases) might deviate from a Brønsted plot. Electron-withdrawing substituents like CN would deviate in the opposite
direction from electron-donating (by resonance) substituents
like OCH_3. The resulting plot then could be S-shaped.

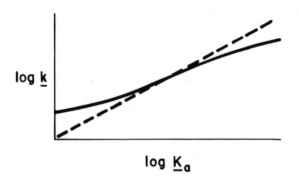

In the construction of a Brønsted plot, extreme caution should
be used to maintain a structurally homogeneous series of
acids (or bases) with the same charge type. Even so,
curvature may result from electrostatic interactions.

References: A. J. Kresge and Y. Chiang, J. Am. Chem. Soc.,
95, 803 (1973); A. J. Kresge, Chem. Soc. Rev., 2, 497 (1973).

3. a. The rate of appearance of enol is given by

$$\frac{d[E]}{dt} = k_2[E^-][HOAc],$$

in which E represents the enol and E^- the enolate anion. The
enolate concentration can be removed by substituting the
expression for the equilibrium constant of step 1 (S stands for

$$K_1 = \frac{[E^-]}{[S][^-OH]}$$

the substrate acetone, and the $[H_2O]$ is included in the constant). Substitution gives

$$\frac{d[E]}{dt} = K_1 k_2[S][^-OH][HOAc].$$

The base hydrolysis constant for acetate is

$$\underline{K}_b = \frac{[^-OH][HOAc]}{[^-OAc]}.$$

Substitution into the rate expression gives

$$\frac{d[E]}{dt} = \underline{K}_1\underline{K}_b\underline{k}_2[S][^-OAc]$$

or

$$= \underline{k}_{-OAc}[S][^-OAc].$$

Since the rate is directly proportional to the concentration of the general base ^-OAc rather than the lyate ion ^-OH, the mechanism is an example of general base catalysis (Type II).

Type II General Base Catalysis

$$\underset{CH_3\overset{\displaystyle O}{\overset{\|}{C}}CH_3}{} + \ ^-OH \ \underset{\Longleftarrow}{\overset{K_1}{\Longrightarrow}} \ ^-CH_2\overset{\displaystyle O}{\overset{\|}{C}}CH_3 + H_2O$$

$$\underset{CH_2=\overset{\displaystyle O^-}{\overset{|}{C}}CH_3}{} + HOAc \ \xrightarrow{k_2} \ CH_2=\overset{\displaystyle OH}{\overset{|}{C}}CH_3 + \ ^-OAc$$

This mechanism is kinetically indistinguishable from rate-determining attack by the general base (Type I), except by means of isotope effects (see part (b)).

Type I General Base Catalysis

$$\underset{CH_3\overset{\displaystyle O}{\overset{\|}{C}}CH_3}{} + \ ^-OAc \ \xrightarrow{k_1} \ ^-CH_2\overset{\displaystyle O}{\overset{\|}{C}}CH_3 \ \longleftrightarrow \ CH_2=\overset{\displaystyle O^-}{\overset{|}{C}}CH_3$$

$$\underset{CH_2=\overset{\displaystyle O^-}{\overset{|}{C}}CH_3}{} + H_2O \ \xrightarrow{fast} \ CH_2=\overset{\displaystyle OH}{\overset{|}{C}}CH_3$$

See problem 6-2 for analogous general acid catalysis mechanisms.

b. The Dawson-Spivey third order term involves simultaneous catalysis by both acid, e.g., HOAc, and base, e.g., ^-OAc.

(i) The large primary isotope effect indicates convincingly that the C-H bond is being broken in the transition state. Such isotope effects are also observed for the acid-catalyzed ($\underline{k}_{HOAc}$, $\underline{k}_{H_3O^+}$) and base-catalyzed

$(\underline{k}^-_{OAc}, \underline{k}^-_{OH})$ terms (problem 6-2). These observations suggest a Type II mechanism for acid catalysis (rapid protonation followed by rate-determining proton abstraction by the conjugate base of the general acid) and direct, rate-determining attack (Type I) by the general base for base catalysis (see part (a)). For the third-order term, the $\underline{k}_H/\underline{k}_D$ implies that the general base is involved in C-H abstraction in the transition state but says nothing about the role of the general acid.

(ii)   The large solvent isotope effect implies that an O-H bond is broken in the transition state. Since all hydroxyl groups are deuterated in D_2O, the O-H (O-D) bond that is broken is presumably that in the catalyzing species, which is HOAc (DOAc) in the third order term. In order to satisfy the conditions of both C-H and O-H bond cleavage in the rate-determining step, the transition state for bifunctional catalysis must involve two concerted proton transfers. These

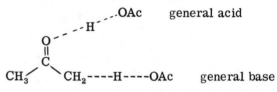

observations contrast with those for the second order acid- and base-catalyzed terms, for which $\underline{k}_{H_2O}/\underline{k}_{D_2O} \sim 1$. The transfer of the proton from H_3O^+ to the carbonyl oxygen does not occur in the rate-determining step, as in the Type II acid catalysis mechanism above. These results suggest that there is no unified mechanism for all acidic and basic species and that the concerted, bifunctional catalysis in the third order term is distinct from the mechanisms of catalysis in the second order term.

(iii)   The Brønsted β is something of a red herring. Normally a very low β is characteristic of specific acid catalysis in a reactant-like transition state, i.e., little dependence on the structure of the general acid. In the present bifunctional case, the action of the general acid is opposite to that of the general base, so the effects tend to cancel. An increase in the effectiveness of the catalyzing base is accompanied by a decrease in the effectiveness of the catalyzing acid. The β_B is in the vicinity of 0.65 and the α_A in the vicinity of 0.5 to give the observed β_{AB} (the difference) of 0.15. This α_A for the third order term is much larger than that ordinarily observed for the second order term with general acid catalysis (0.2). The difference is in accord with a bifunctional transition state for the third order mechanism with considerably more proton transfer to the carbonyl oxygen than in the monofunctional (second order) case.

Reference: A. F. Hegarty and W. P. Jencks, J. Am. Chem. Soc., 97, 7188 (1975).

4. Of the two types of general acid catalysis (I and II, see problem 6-2), only the first is in accord with the solvent isotope effect.

(I) $S + H-A \xrightarrow{\text{slow}} \overset{+}{S}-H + A^-$

(II) $S + H_3O^+ \rightleftharpoons \overset{+}{S}H$

         $^+SH + A^- \xrightarrow{\text{slow}}$ products

The large k_{H_2O}/k_{D_2O} is consistent with the rate-determining cleavage of the H-O bond of the general acid. Therefore it is likely that a proton is transferred directly to the ynamine in the rate-determining step, and all subsequent steps are fast. Similar

$ArC\equiv CNR_2 + H-A \xrightarrow{\text{slow}} Ar\overset{H}{C}=\overset{+}{C}NR_2 + A^- \xrightarrow[H_2O]{\text{fast}}$

$Ar\overset{+}{C}H=CNR_2 \xrightarrow[-H^+]{\text{fast}} ArCH=\overset{OH}{C}NR_2 \xrightarrow{\text{fast}} ArCH_2\overset{O}{C}NR_2$

mechanisms have been proposed for the acid-catalyzed hydration of 1-alkynyl ethers and thioethers. The rate-determining proton transfer is consistent with the large negative $\Delta \underline{S}^{\ddagger}$, since the bimolecular transition state has considerable separation of charge ($ArCH=\overset{+}{C}NR_2 + A^-$). The two components of the reaction (ynamine, general acid) and the solvent suffer loss of translational degrees of freedom. The ρ^* is negative because of the development of positive charge in the transition state on the carbon atom near the R substituents.

Reference: W. F. Verhelst and W. Drenth, <u>J</u>. <u>Am</u>. <u>Chem</u>. <u>Soc</u>., 96, 6692 (1974).

5. The k_{H_2O}/k_{D_2O} is less than unity, so the first step (protonation) in either mechanism cannot be rate limiting. The inverse isotope effect is characteristic of a preequilibrium protonation, since D_3O^+ is a stronger acid than H_3O^+. The observation of specific acid catalysis also demonstrates that the first step cannot be rate limiting (slow protonation, S + HA, would have given Type I general acid catalysis (see problem 6-2)). Consequently, for either mechanism, the second step, nucleophilic attack by water on the protonated substrate, must be rate limiting. Since $k_{obs} = K_1 k_2$, the $\rho^*_{obs} = \rho_1^* + \rho_2^*$. For either mechanism, ρ_1^* is negative (electron-withdrawing groups favor the left side) and ρ_2^* is positive (positive charge is moved away from the R' groups in the transition state). In mechanism (a), <u>O</u>-protonation, the positive charge is not close to the immediate vicinity of the R' groups, so ρ_1^* and ρ_2^* should both be relatively small. In mechanism (b), <u>N</u>-protonation, the nitrogen atom to which the R' groups are bonded bears the positive charge, so ρ_1^* and ρ_2^* should be large. The reasonably large ρ^* of -1.0 is consistent with <u>N</u>-protonation, but the result is not definitive because of the canceling effect of ρ_1^* and ρ_2^*. The two mechanisms should be more clearly distinguished by the effect of incorporation of a small ring as the R groups. In mechanism (a), a four-membered ring can be comfortably accommodated between an axial and an equatorial position in the pentacoordinate intermediate (A). The

A

B

normal 90° angle in A (and in the transition state leading to it) is less strained than the angle in the starting material, so the small ring should cause a rate acceleration. In the transition state to direct displacement (B) in mechanism (b), the ring angle approaches 120° and is larger and more strained than that in the starting material, so the small ring should cause a rate deceleration. The observed deceleration is consistent with mechanism (b), $\underline{N}$-protonation/direct displacement. No intermediate need be invoked, so the mechanism is closely rated to the classic $S_N 2$ displacement on carbon. In contrast to these phosphinamides, carboxylic amides appear to protonate predominantly on oxygen at high acidity. Protonation occurs more readily on the nitrogen of phosphinamides because amide-like overlap between the $2p$ nitrogen lone pair and the $3p$ phosphorus orbitals ($-\ddot{P}-NR_2' \longleftrightarrow P = \overset{+}{N}R_2'$) is less important than the $2p$-$2p$ amide overlap ($-\overset{O}{C}-NR_2' \longleftrightarrow -\overset{O}{C}=\overset{+}{N}R_2'$). Protonation ($-\overset{O}{P}-\overset{+}{N}HR_2'$, $-\overset{O}{C}-\overset{+}{N}HR_2'$) destroys this overlap.

Reference: T. Koizumi and P. Haake, <u>J</u>. <u>Am</u>. <u>Chem</u>. <u>Soc</u>., <u>95</u>, 8073 (1973).

6. a. The enolization/halogenation mechanism applies (see problem 10-3).

This mechanism is Type I general base catalysis ($A_{\underline{i}}^{-}$ is the general base). It should exhibit a large kinetic isotope effect, since the C-H bond is broken in the rate-limiting step, and it is zero order in $[I_2]$, which enters after the rate-limiting step

b. The rate-determining step involves development of a negative charge. The polar (inductive/field) effect of the positive charge on the methylated nitrogen should be a stabilizing influence. In the 2-substituted case (A), the electrostatic effect is quite large (C/A = 1140) because the acetyl side chain is close to the positive charge (E). In the 4-substituted

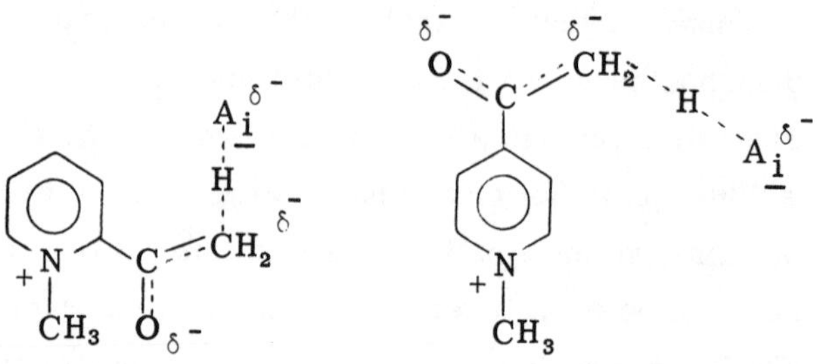

E   F

case (B), the electrostatic effect is much smaller (D/B = 14) because of the large distance between the acetyl group and the N-methyl group (F). A similar effect is noticed on protonation. Protonation of A (to give G) causes a 200-fold

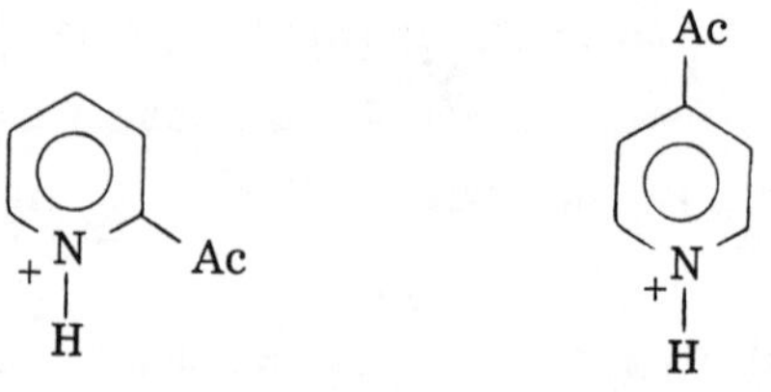

G   H

acceleration in the iodination rate, whereas protonation of B (to give H) causes only a 20-fold acceleration.

c. The metal-catalyzed reaction of 2-acetylpyridine (A) must follow essentially the same mechanism as in (a), but with some kind of transition state stabilization through complexation. Chelation of the metal to the nitrogen atom and the enolate oxygen, as in J, should provide an electrostatic stabilization that is similar to but more effective than

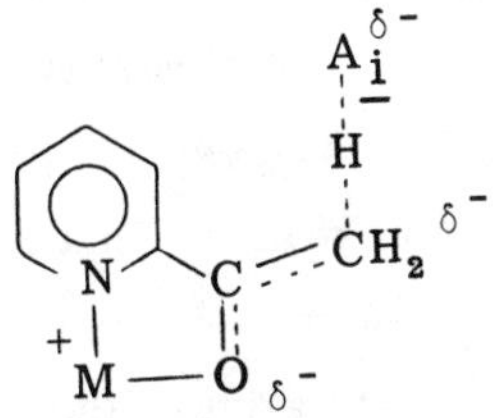

N-methylation or N-protonation (see (b)), since the positive charge on the metal M is larger and closer to the site of the negative charge development. Such chelation is not possible when the acetyl group is in the 4 position. For A, complexation with the metal would be treated kinetically as a rapid preequilibrium prior to the rate-determining proton abstraction.

Reference: B. G. Cox, J. Am. Chem. Soc., 96, 6823 (1974).

7. a. Because the reaction is specific acid catalyzed, the first step must be a rapid protonation by H_3O^+, and the functional nucleophile in the second step must be H_2O rather than the general base. The inverse isotope effect confirms that there is a rapid (not rate-determining) protonation step (D_3O^+ is a stronger acid than H_3O^+). The following mechanism is in agreement with the observations.

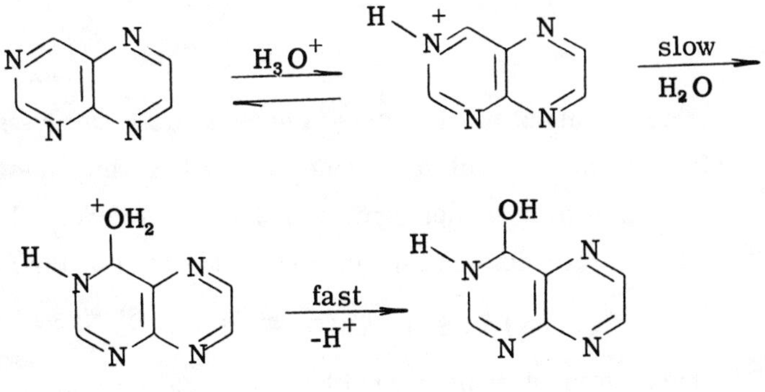

b. The large primary isotope effect indicates that an O-H bond is broken in the transition state. Thus the rate-determining step must include a proton transfer from H_2O to nitrogen, $H-O----H----N$. The abnormally large catalytic coefficient for H_2O, however, suggests that water, in addition to being a general acid, must serve in another capacity. It is likely that a second water molecule in the transition state serves as the nucleophile that delivers the hydroxyl group. This phenomenon is called bifunctional catalysis (also see problem 10-3b) and is best represented by the transition state A. Other catalysts with two oxygen atoms ($H_2PO_4^-$, HCO_3^-, $H_2AsO_4^-$) are

A

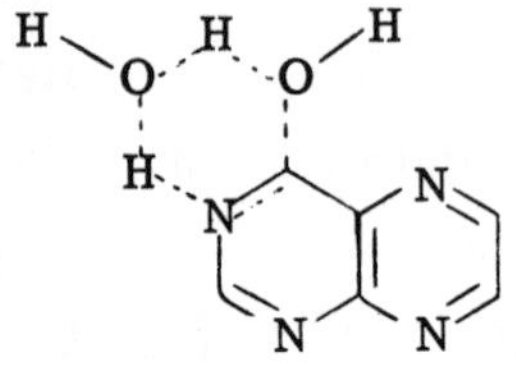

particularly effective in catalyzing the hydration reaction and may be bifunctional. In these cases the catalysts deliver the proton and lead in the nucleophile, as in B.

B

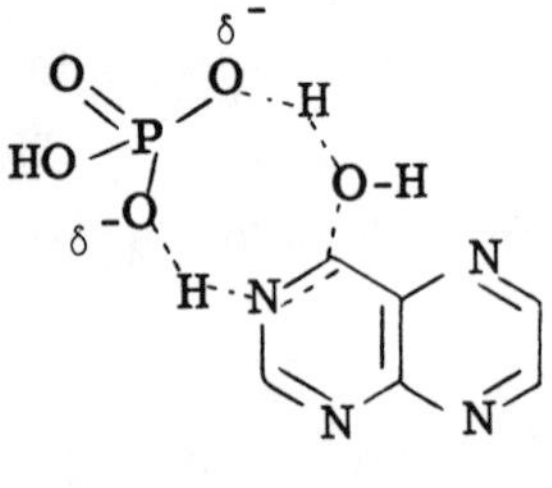

c. The two usual mechanisms of general base catalysis involve rapid reaction with hydroxide followed by rate-determining reaction with the conjugate acid HA of the general base (Type II) and rate-determining proton abstraction by the general

$$\text{(Type II)} \quad S + {}^-OH \rightleftharpoons {}^-S-OH \xrightarrow[\text{HA}]{\text{slow}} SH-OH + A^-$$

base as must be envisaged here with an intervening water molecule (Type I) (see problem 10-3). Mechanisms are excluded in which A^- serves as a nucleophile rather than as a

(Type I) $S + H-O-H + A^- \xrightarrow{\text{slow}} S^- -OH + HA$

$\xrightarrow{\text{fast}} SH-OH + A^-$

base. Mechanisms (I) and (II) are kinetically indistinguishable. The k_{H_2O}/k_{D_2O} for Type I should be greater than unity because of the proton transfer. In mechanism II a normal isotope effect in the second step should be offset by an inverse isotope effect in the first step (^-OD is a better nucleophile and should favor the right side). If the proton transfer is not well developed in the transition state of the second step, the observed k_{H_2O}/k_{D_2O} would be in accord with this mechanism. The authors prefer to combine the two events of (II) in a single, kinetically equivalent step, with the transition state C. Again, the assumption must be made that the proton is not

C

appreciably transferred. The available data do not define the mechanism unambiguously in the high pH region.

Reference: Y. Pocker, D. Bjorkquist, W. Schaffer, and C. Henderson, J. Am. Chem. Soc., 97, 5540 (1975).

8. a. The A-S_E2 mechanism will show general acid catalysis since the proton that is transferred in the transition state must still be part of the general acid. The A-1 mechanism will show specific acid catalysis, since the proton is introduced in a fast step (therefore not Type I general acid catalysis (problem 6-2)) and A^- (the conjugate base of the general acid) does not enter elsewhere into the mechanism (therefore not Type II).

   b. The plots show a linear dependence of the rate on the concentration of buffer, corrected to constant $[H^+]$. This behavior is characteristic of general acid catalysis and is therefore consistent with the $A-S_E2$ mechanism for hydration (direct, rate-limiting protonation of the alkene).

c. As the buffer concentration is changed, the ionic strength will also vary, and this variation will affect $[H^+]$. If a high ionic strength is used to wash out this salt effect, specific ionic interactions can entirely mask general acid catalysis. Although the hydration of functionally substituted alkenes (styrenes, vinyl ethers, enamines, ketene acetals) had previously been found to be subject to general acid catalysis and hence have the $A-S_E2$ mechanism, aliphatic alkenes had not been observed to have these properties. These latter results must have been misinterpreted because of high buffer concentrations and hence high ionic strengths. The present authors used ionic strengths of no more than 0.1 M, at which concentrations activity coefficients can be reliably calculated from the Debye-Hückel equation. Since the rate is proportional to $[H^+]$, as well as to $[HA]$, the effect of the buffer must be obtained by correcting for the $[H^+]$ term as $[HA]$ is varied. To make this correction, the authors measured $\underline{k}_{H^+}$ in dilute $HClO_4$ solutions, and then adjusted all the rate to the $[H^+]$ of the most concentrated buffer. The circle points in the plots in (b) represent the residual effect of buffer concentration, $[HA]$, after the effect of $[H^+]$ is adjusted to a constant value. The residual linear dependence of the rate on $[HA]$ substantiates the existence of general acid catalysis.

Reference: A. J. Kresge, Y. Chiang, P. H. Fitzgerald, R. S. McDonald, and G. H. Schmid, J. Am. Chem. Soc., 93, 4907 (1971).

9. a. Dependence of the rate on $\underline{h}_0$ indicates that the transition state contains only the substrate and a proton, according to the Zucker-Hammett hypothesis. Therefore the rate-determining step is the unimolecular breakdown of the conjugate acid of the

substrate. The molecule of water needed to complete the
hydrolysis enters after the slow step.

b. Proportionality of the rate with $[H^+]$, by the Zucker-Hammett
hypothesis, suggests that the transition state contains substrate, a proton, and a molecule of water. Thus the rate-
determining step is attack of water on the conjugate acid of
the substrate.

It is noted that in the two mechanisms, (a) and (b), different
conjugate acids are used. As is normally the case, all
possible conjugate acids form, according to their respective
pK_a's, but we write only the one that proceeds to the favored
reaction for a given substrate. Protonation of the ring oxygen in β-propiolactone gives a good leaving group for opening
of the strained four-membered ring. This driving force is
not so prevalent in the five-membered γ-butyrolactone, so
the reaction proceeds more normally by protonation of the
carbonyl oxygen.

c. The reverse of the γ-butyrolactone hydrolysis involves protonation of the hydroxy acid, several fast steps, slow loss of H_2O to give the conjugate acid of the lactone, and finally loss of the catalytic proton (the reverse of the mechanism in (b)). Thus the transition state contains only the substrate and a single proton (the water molecule is in the process of being lost, not gained). According to the Zucker-Hammett hypothesis, the lactonization rate should then be proportional to $\underline{h}_0$. This result is general. In these reactions involving substrate, a proton, and H_2O, if the reaction in one direction follows $[H^+]$, the reverse reaction must follow $\underline{h}_0$, according to the hypothesis. Unfortunately, there are examples when both directions of the reaction follow $\underline{h}_0$. To prove that a reaction with only the substrate and a proton in the transition state follows $\underline{h}_0$ requires the assumptions that the substrate be a Hammett base (if not, another, more appropriate $\underline{h}$ function can be substituted) and that the activity coefficient of the transition state be close to that of the protonated substrate ($\gamma_{\ddagger} \sim \gamma_{HS^+}$). These assumptions are quite frequently met. To prove that a reaction with the substrate, a proton, and H_2O in the transition state follows $[H^+]$ requires more arbitrary assumptions, such as that the activity coefficient of the transition state be close to the product of those for the substrate and the hydrogen ion ($\gamma_{\ddagger} \sim \gamma_S \gamma_{H^+}$). Although this requirement is sometimes met, its breakdown leads to the result that the rate can be proportional to $\underline{h}_0$ or $[H^+]$, or some intermediate dependence. For these reasons, the Zucker-Hammett approach is not entirely reliable and must be used with full cognizance of its limitations.

Reference: K. B. Wiberg, "Physical Organic Chemistry," Wiley, New York, NY, 1964, pp 430-34.

10. The concave upwards Hammett plot is indicative of a change in mechanism (competing pathways). For electron-donating substituents such as $\underline{p}$-CH_3O ($\sigma^+ < 0$), ρ is large and negative, so the transition state must be electron deficient relative to the ground

state. The A-S_E2 mechanism (rate-determining proton transfer)
brings positive charge closer to the Ar group and therefore is in
accord with these observations. The large isotope effect (2.50)
is consistent with H-O bond cleavage in the transition state. The
proportionality of the rate with $\underline{H}_0$ is characteristic of the rate-
determining unimolecular breakdown of the protonated substrate
(see problem 10-9). Thus all the evidence points toward the A-
S_E2 mechanism for the $\underline{p}$-CH_3O substrate throughout the entire
range of acidity. The $\rho = 0$ for electron-withdrawing substituents
such as $\underline{p}$-NO_2 ($\sigma^+ > 0$) is characteristic of the normal acid-
catalyzed ester hydrolysis mechanism (tetrahedral intermediate).
The positive ρ for the first step (protonation equilibrium) cancels
the negative ρ for the second (attack by H_2O) (see problem 9-8).
This ρ was measured at low acidity, and the inverse isotope
effect under these conditions is in accord with a rapid protonation (D_3O^+ being a stronger acid than H_3O^+) followed by the rate-
determining reaction with H_2O. As the acidity increases, the
mechanism must change to the A-S_E2 (rate-determining proton
transfer), since the k_{H_2O}/k_{D_2O} increases to 3.25. The curvature
in the $\underline{H}_0$ plot could be the result either of this change in mechanism or of an intermediate $\underline{h}_0/[H^+]$ behavior (see problem 10-9).
The authors' analysis of the curve indicates that log $\underline{k}$ is linear
with $\underline{H}_0$ at high acidity (A-S_E2 region, in agreement with the $\underline{p}$-
CH_3O result) but is clearly curved at low acidity (normal ester
mechanism region, in agreement with results with regular esters,
such as isopropyl acetate, which give a curved log $\underline{k}$ vs. $\underline{H}_0$
profile). The A-S_E2 mechanism is favored by high acidity and by
electron-donating substituents. It is reasonable that electron
donation would accelerate the A-S_E2 pathway (and electron withdrawal disfavor it), since a carbonium ion is formed. At higher
acidities, the A-S_E2 mechanism takes over from the normal ester
mechanism even for electron-withdrawing groups because of the
increased scarcity of H_2O.

Reference: D. S. Noyce and R. M. Pollack, J. Am. Chem. Soc., 91, 119 (1969).

11. The decomposition of the intermediate provides an excellent measure of the relative leaving abilities of the substituted pyridines vs. the phenolate ion, as reflected in the relative values of k_{-1} and k_2. The identity of the rate-determining step depends on the ratio of these two quantities. Thus when $k_{-1} \gg k_2$ (pyridine is the better leaving group), the k_{obs} becomes $K_1 k_2$ ($K_1 = k_1/k_{-1}$) and the second step is rate determining. When $k_2 \gg k_{-1}$ (phenolate is the better leaving group), the k_{obs} becomes k_1 and the first step is rate determining. The curvature in the Brønsted plot appears to result from a change in the rate-determining step. The same phenolate ion is used throughout, but the pyridine substituent is variable. For electron-withdrawing substituents such as 4-CN, the conjugate acid of the pyridine is strongly acidic ($pK_a = 1.98$ for 4-CN) so the neutral pyridine is a weak base and a good leaving group (the Brønsted slope $\beta = 0.9$ in this region). For these substituents, the second step is rate determining. For electron-donating substituents such as 4-N(CH$_3$)$_2$, the conjugate acid is much weaker ($pK_a = 9.55$) so the neutral pyridine is a stronger base and a worse leaving group. For these substituents, the first step is rate determining. The slope β is 0.2 in this region. At the point of maximal curvature ($pK_a = 7.8$), the rate of breakdown of the tetrahedral intermediate should be similar in the two directions ($k_{-1} \sim k_2$), i.e., the leaving group abilities are about the same. The break occurs somewhere between 4-methyl- and 4-aminopyridine. Since the pK_a of the conjugate acid of the phenolate is about 4.0 and that of the hypothetical pyridine of equivalent leaving ability is 7.8, one can conclude that pyridines are about 10^4 better leaving groups than phenolates, at constant basicity.

Reference: E. A. Castro and F. J. Gil, J. Am. Chem. Soc., 99, 7611 (1977).

12. The solvation parameter ϕ (or $\phi_{\ddagger}$) is a measure of the interaction of the solvent with the cation. The equilibrium ϕ values range from strongly interacting oxonium ions ($\phi \sim 0.7$) to moderately interacting ammonium ions ($\phi \sim 0.0$; $\phi = 0.0$ for Hammett bases) and weakly interacting carbocations ($\phi \sim -1.0$). The negative values of $\phi_{\ddagger}$ in the present cases are consistent with the weak interactions of the ground state carbocations. The comparative values of $\phi_{\ddagger}$ do not support significant differences in transition state solvation between protonated alkenes and alkynes, since the $\phi_{\ddagger}$ are very similar for analogous pairs, usually less than 0.1. The larger negative values of $\phi_{\ddagger}$ as the aryl substituent becomes more electron donating are consistent with weaker solvent interactions as delocalization increases. With the elimination of transition state solvation differences as an explanation for the unusual similarity in alkene and alkyne reactivity toward protonation, another factor must be sought. The authors suggest ground state thermochemical differences (greater relative strengths of double bonds) coupled with different stabilities of halogen-bridged double and triple bonds. Certainly the protonation and halogenation mechanisms are sufficiently different that strong parallelism between relative rates is not necessary.

References: V. Lucchini, G. Modena, G. Scorrano, and U. Tonellato, J. Am. Chem. Soc., 99, 3387 (1977); G. Modena, F. Rivetti, G. Scorrano, and U. Tonellato, ibid., 99, 3392 (1977).

11
ORBITAL SYMMETRY

1. Using (a) the Woodward-Hoffman HOMO approach and (b) the
   Evans-Dewar aromatic transition state method, demonstrate
   why the thermally allowed ground state [3,3] sigmatropic shift
   has the supra-supra or antara-antara stereochemistry. Illustrate your answer with carefully drawn molecular orbitals.
   Use the Cope rearrangement of 1,5-hexadiene as your example.

2. Consider the following reactions (p. 220).

   a. Give the pericyclic nomenclature for the two reactions.

   b. Reaction A (1,4 elimination) is much faster ($\Delta \underline{G}^{\ddagger} = 28.5$ kcal
      mol^{-1}, reaction at 100-120 °C) than reaction B (1,6 elimina-
      tion, $\Delta \underline{G}^{\ddagger} = 39.5$ kcal mol^{-1}, reaction at 240-275 °C). Suggest
      an explanation in terms of the principles of orbital symmetry.

A $\quad$ SO$_2$ $\quad\xrightarrow{\Delta}\quad$ $+$ $\quad$ SO$_2$

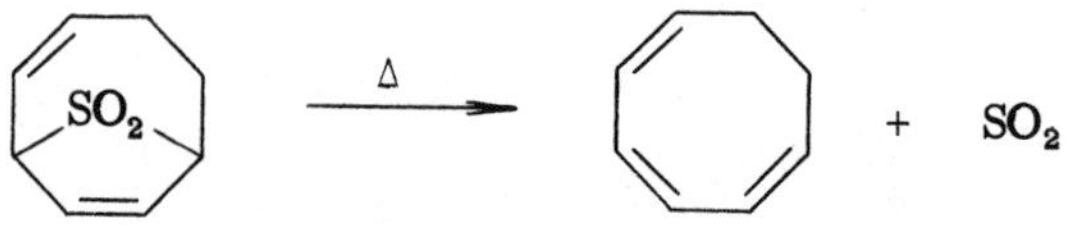

B $\quad$ SO$_2$ $\quad\xrightarrow{\Delta}\quad$ $+$ $\quad$ SO$_2$

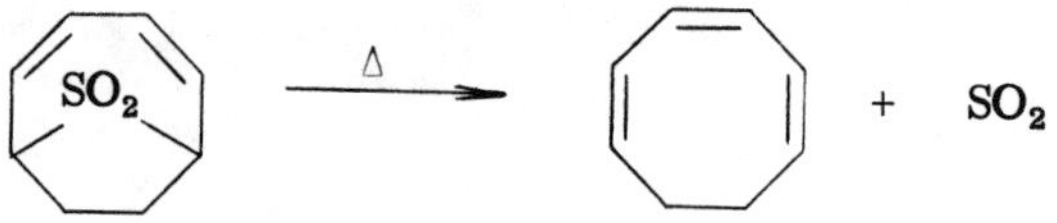

c. For reaction A, $\Delta\underline{H}^{\ddagger} = 31.2$ kcal mol^{-1} and $\Delta\underline{S}^{\ddagger} = +7$ eu. For reaction B, $\Delta\underline{H}^{\ddagger} = 30.0$ kcal mol^{-1} and $\Delta\underline{S}^{\ddagger} = -18$ eu. How would these results influence your answer to (b)?

3. Propargyl esters may be converted thermally to α, β-unsaturated ketones.

$$HC\equiv CCR'R''O\overset{O}{\overset{\|}{C}}R''' \longrightarrow R'R''C=CHC\overset{O}{\overset{\|}{}}R'''$$

a. The mechanism involves four steps, a Claisen/Cope-type reaction, a 1,3-acyl shift, a cyclic decarbonylation, and an enol-keto isomerization. Draw out the mechanism, showing each step.

b. Label each step by its pericyclic nomenclature, and indicate whether (and why) it is allowed or forbidden by orbital symmetry.

4. A by-product from the thermal reaction of ethynylbenzene and 3,4-diphenylthiophene 1,1-dioxide is a naphthalene. Two different modes of deuterium labeling gave the following results.

a. Draw out the mechanism of the reaction.

b. Give the full pericyclic nomenclature for each step.

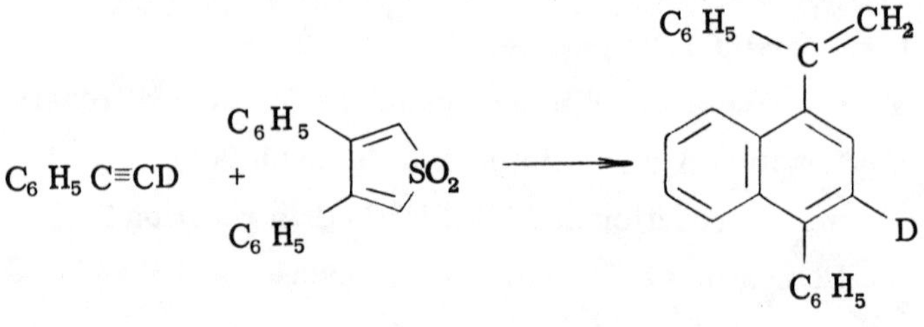

C$_6$H$_5$C$\equiv$CD $\quad + \quad$ [thiophene dioxide with C$_6$H$_5$, C$_6$H$_5$, SO$_2$] $\quad \longrightarrow \quad$ [naphthalene product with C$_6$H$_5$, =CH$_2$, C, D, C$_6$H$_5$]

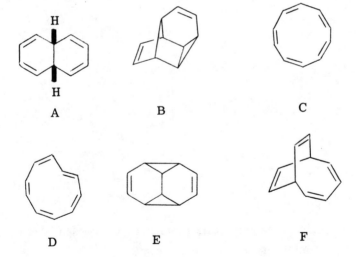

5. a. Which pairs of the following $C_{10}H_{10}$ hydrocarbons might be thermally interconverted (orbital symmetry allowed)? Label each reaction you find by its pericyclic nomenclature. Ignore antarafacial possibilities and higher order cycloadditions $(2 + 2 + 2)$.

b. Which pairs might be photochemically interconverted? Use the same ground rules as in (a).

6. <u>Related subject:</u> <u>introduction to kinetics</u>

Compound A is thermally stable in the sense that heating does

A

not produce isomerization. Reaction of A with tetracyanoethylene
(TCNE) produces a 1:1 adduct, presumably via a pericyclic
reaction. The rate of adduct formation is independent of the
concentration of TCNE, but dependent on the concentration of A
to the first power.

a. Draw the steps in the formation of the adduct, showing its
    structure and stereochemistry.

b. List the rate constants of the various steps in order of
    increasing magnitude. Explain your order.

7. Quadricyclane can react with electron rich alkenes by either of
    two modes, to form A or B.

a. What is the pericyclic nomenclature for each reaction?
b. Which mode is favored and why?
c. Norbornadiene has two analogous modes of cycloaddition.
    Write out the two modes and answer the questions given in
    (a) and (b). Compare the two results.

8. The structurally related compounds A and B rearrange at 400 °C to rather different products. Draw out the pericyclic steps for these reactions and suggest reasons for the different pathways.

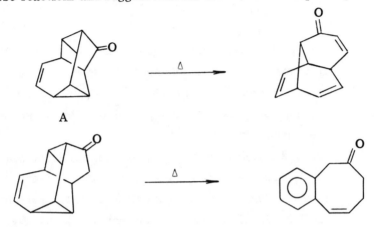

A

B

9. Related subject: introduction to kinetics

Pyrolysis of homobasketene (A) at 180 °C gives benzene and cyclopentadiene. When a 20-fold excess of maleic anhydride is included, the product is B.

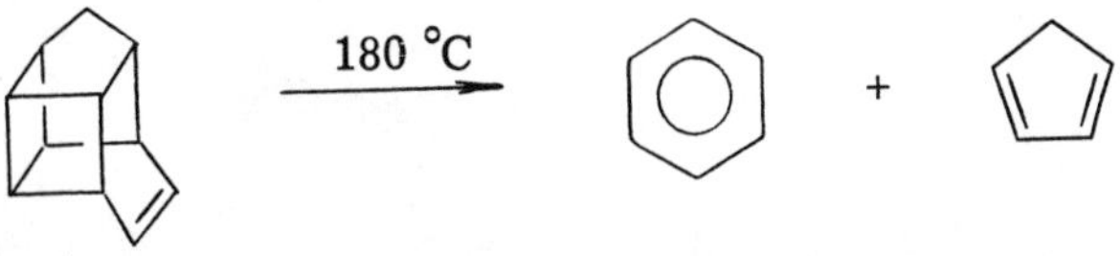

a. Draw out the steps for both reactions. Base your mechanism for the first reaction on the trapping results of the second reaction. Note that B is a rearrangement product of the primary trapping product.

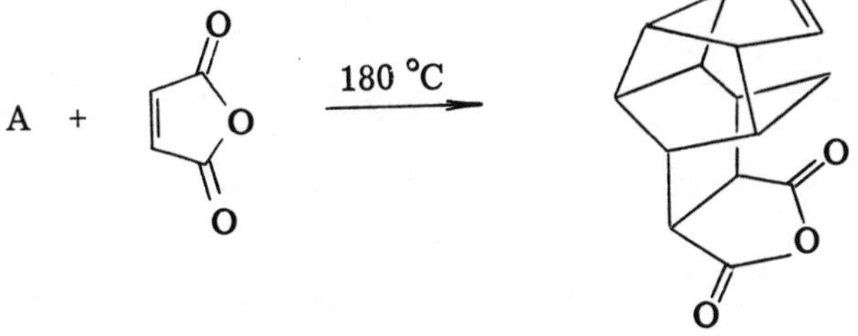

b. In a trapping reaction of this sort (C is benzene plus

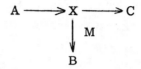

cyclopentadiene), what will be the effect on the overall rate from the addition of maleic anhydride (M)?

10. a. Draw out three pericyclic mechanisms (several steps are possible) whereby cyclooctatetraene (COT) can automerize, i.e., isomerize into itself.

b. The 1,2-dimethylcyclooctatetraene initially isomerizes to the 1,3 and 1,4 isomers, and some of the 1,5 isomer later is formed. The 1,5 isomer goes first to the 1,4 and later to the 1,3. Discuss each of the mechanisms in (a) in terms of these results. Which mechanism(s) is (are) favored?

SOLUTIONS

1. a. As presented in their earliest papers, Woodward and Hoffman
   suggested that the highest occupied molecular orbital (HOMO)
   be examined for the overlap between orbitals with like sign in
   the transition state in order to ascertain whether a given pathway is allowed. Although the use of correlation and state
   diagrams is more general, organic chemists continue to use
   the original HOMO method as a useful mnemonic device. For
   the [3,3] sigmatropic shift, the transition state can be considered to be a pair of allyl fragments. The allylic HOMO
   has a node at the central carbon atom. Two such fragments

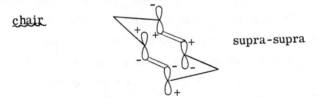

can be brought together in chair, boat, twist-boat, or half-
chair conformations. The chair form gives a fully bonding

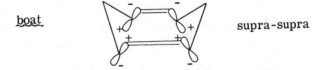

supra-supra overlap for an allowed reaction. The boat form

similarly gives an allowed supra-supra arrangement, but is
frequently disfavored with respect to the chair form, possibly
because of repulsive interactions between the central carbons.
The twist-boat form gives a relatively strained but allowed
antara-antara arrangement. The half-chair form would give

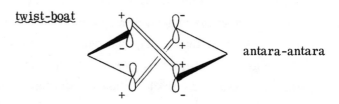

twist-boat   antara-antara

the supra-antara result, but the reaction is forbidden because one overlap is between orbitals of unlike sign.

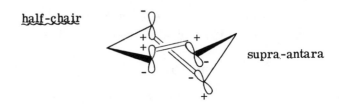

half-chair   supra-antara

b. To apply the Evans-Dewar method, one must count the number of electrons in the pericyclic ribbon and the number of nodes in the transition state. The ribbon consists of the π electron system plus the σ bonds that are being made or broken. The present case contains six electrons (four π and two σ electrons in the starting material ribbon or three π electrons in each of the two allyl fragments in the transition state). For such Hückel systems (4n + 2 electrons) the reaction is allowed when there are an even number (0, 2, 4...) of nodes. For 4n or Möbius systems, the reaction is allowed when there are an odd number of nodes. The transition states have been drawn in part (a). Each allyl fragment always has one node at the central carbon. The chair and boat forms have no additional nodes at the points of σ bond making or breaking, so the total number of nodes (2) is even and the supra-supra arrangement is allowed for this Hückel system (six electrons). Similarly, the twist-boat form has two nodes and six electrons (Hückel), so the antara-antara arrangement is also allowed.

The half-chair has an extra node (sign inversion) at one σ bond portion of the ribbon, so the total number (3) is odd and the supra-antara arrangement is forbidden for this six electron system. The Woodward-Hoffman HOMO and the Evans-Dewar aromatic transition state method give identical results in this case. They are the most convenient procedures for determining the allowedness of pericyclic reactions. The choice between the two is often a matter of personal preference, but the Evans-Dewar procedure is faster in a number of instances.

Reference: T. L. Gilchrist and R.C. Storr, "Organic Reactions and Orbital Symmetry," Cambridge Univ. Press, Cambridge, England, 1972, pp 27-31, 38-43, 220-23.

2. a. These extrusion processes are termed cheletropic reactions. In order to form cis double bonds (required by the size and unsaturation of the rings), both reactions must be disrotatory in the hydrocarbon portion. According to the cheletropic selection rules for disrotatory processes, the four π electron system in A (six total electrons) must involve a linear mode, whereas the six π electron system in B (eight total electrons) must involve a nonlinear mode. As written, A is a retro $_\pi 4_s + _\omega 2_s$ reaction, and B is a $_\pi 6_s + _\omega 2_a$ reaction. The subscripts signify suprafacial (s) and antarafacial (a), respectively. The π and the ω signify the destruction of a pi bond and a lone pair, respectively. The 4 + 2 reaction can be linear and supra-supra in the disrotatory mode, but the 6 + 2 reaction must be nonlinear and supra-antara (or antara-supra) to be disrotatory.

    b. The requirement that the reaction be disrotatory in order to produce cis double bonds is consistent with a stereochemically facile supra-supra, linear pathway for A. This same (linear) pathway for B would have to be conrotatory and hence lead to a trans double bond in the product. The observed cis stereochemistry in reaction B therefore is consistent with a stereochemically somewhat infacile (but allowed) nonlinear

pathway. The stereochemical constraints raise the free
energy of activation for B with respect to that of A.

c. If both reactions are concerted, it seems likely (though not
necessary) that the activation entropies would be comparable
Moreover, orbital symmetry differences should be found
mainly in the activation enthalpy. In fact, the reactions are
almost isoenthalpic and the activation entropies are quite
different. It seems safe to conclude that the linear, four π
electron reaction A is concerted, with nearly synchronous
rupture of the two C—S bonds. The six π electron reaction
B may be nonlinear and concerted, or it may be nonsyn-
chronous. Although a choice cannot be made unambiguously
on the basis of the available evidence for reaction B, it can
be concluded that there is no synchronous, concerted (nonlinear) mechanism for B that is within 10 kcal mol^{-1} in
$\Delta\underline{G}^{\ddagger}$ of the synchronous, concerted (linear) mechanism for
A. The nonlinear, antarafacial mode, though allowed, must
be considerably disfavored on stereochemical grounds.

References: W. L. Mock, J. Am. Chem. Soc., 92, 3807
(1970); 97, 3673 (1975); T. L. Gilchrist and R.C. Storr,
"Organic Reactions and Orbital Symmetry," Cambridge Univ.
Press, Cambridge, England, 1972, pp 85-88, 192-94.

3. a.

b. Reaction A is an allowed, supra-supra sigmatropic shift of
   order [3,3]. It is a variant of the Cope rearrangement, with
   the interesting difference that a triple bond and a carbonyl
   group replace the two double bonds. The [1,3] sigmatropic
   shift of reaction B is allowed either in an antarafacial
   fashion (on the C=C-O portion) with retention (on the aryl
   group) or in a suprafacial fashion with inversion. Although
   the antarafacial pathway is normally stereochemically disfavored, the special geometry of the allene function may
   remove this problem. No choice between the pathways can
   be made, nor for that matter can a stepwise (acyl cleavage)
   mechanism be eliminated. The mechanistic pathway thus
   far (A, B) is supported by the regular observation of α-formyl
   α,β-unsaturated ketones [R'R''C=C(HC=O)(R'''C=O)], the
   suggested product of reaction B. Reaction C is a linear
   cheletropic extrusion of carbon monoxide (six total electrons,
   $_\pi 2_s + _\sigma 2_s + _\sigma 2_s$, or a retro $_\pi 2_s + _\sigma 2_s + _\omega 2_s$) (see problem
   11-2). The subscripts σ, π, and ω indicate whether sigma
   bonds, pi bonds, or lone pairs are being destroyed in the
   given direction of the reaction. The question of disrotation
   (allowed in the linear mode of 6 q total electrons) is irrelevant, since there is no relative stereochemistry for the
   termini of the would-be π electron fragment. Since a linear
   mode is available, the reaction is probably concerted, but

the point cannot be proved. The enol-keto isomerization (D) is another [1,3] sigmatropic shift. An inversion pathway (allowed for a suprafacial shift) does not apply, since a migrating proton is only capable of retention. Thus the reaction is either antarafacial with retention, or (more likely) it is stepwise.

Reference: W. S. Trahanovsky and S. L. Emeis, J. Am. Chem. Soc., 97, 3773 (1975).

4. a. The following reaction sequence can be deduced from the deuterium labeling patterns.

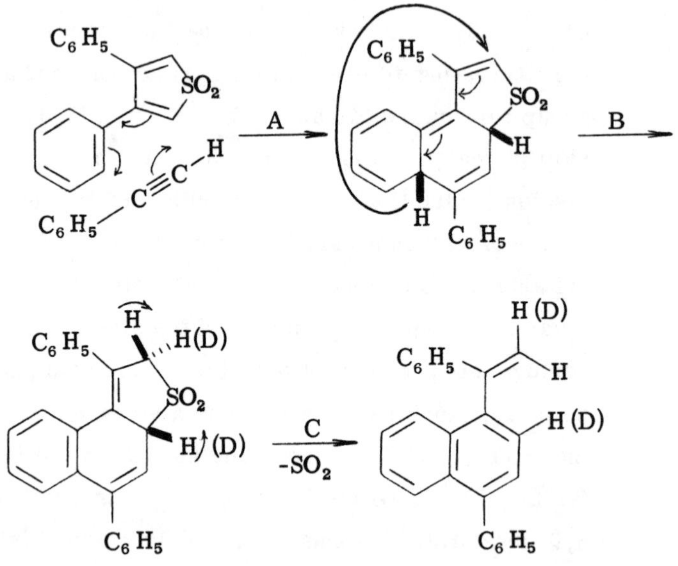

b. The fact that the single deuterium of the C≡CD substrate is attached to the naphthalene ring in the product indicates that a cycloaddition is necessary and that one of the phenyl groups on the thiophene ring must be a party to this cycloaddition. A simple Diels-Alder reaction (A, $\pi^4_s + \pi^2_s$) forms the dihydronaphthalene ring, which is still fused to the thiophene ring. The supra-supra nature of the cycloaddition places the two indicated hydrogens cis to each other. The SO_2 cannot be extruded directly from the product of reaction A, since one of the carbons to which it is attached is unsaturated (both must be saturated). A 1,5-sigmatropic shift puts the lower-most indicated hydrogen on the previously unsaturated car-bon (reaction B). After this suprafacial shift, the migrated hydrogen is still above the plane of the paper, but the hydro-gen (labeled as deuterium) originally bonded to that same carbon is now below the plane of the paper. Although this sigmatropic shift appears to be rather stretched, the expected stereochemistry is demonstrated by the last step. The cheletropic elimination of SO_2 (retro $_\pi 4_s + _\omega 2_s$) gives the observed product. This system generates four π elec-trons (six total electrons, counting the two on SO_2). The favored linear process has the disrotatory stereochemistry. Of the two possible disrotatory modes, that with rotation of the two indicated hydrogens towards each other is preferred in order to give the cis double bond in the naphthalene ring. Rotation of these protons inwards moves the labeled hydro-gen (D) outwards to give the observed stereochemistry of the exocyclic double bond. This reaction sequence involves a cycloaddition, a sigmatropic shift, and a cheletropic elimina-tion. Thus, of the four major types of pericyclic reactions, only the electrocyclic ring closure/opening is lacking.

Reference: R. G. Nelb II and J. K. Stille, <u>J</u>. <u>Am.</u> <u>Chem.</u> <u>Soc.</u>, <u>98</u>, 2834 (1976).

5. a. (i)   A $\rightleftharpoons$ B

        An internal Diels-Alder reaction in A ($_\pi 4_s + _\pi 2_s$ cyclo-addition) can occur, with the double bond on one side of the ring adding across the butadiene fragment of the other side. The result is B'; the reader should

B'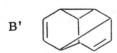

        satisfy himself that B and B' are identical (rotate the page clockwise 120°).

(ii) $$A \rightleftharpoons C$$

A six electron, disrotatory electrocyclic reaction produces the all-cis structure.

A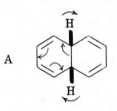

(iii) $$E \rightleftharpoons F$$

Another internal Diels-Alder reaction, this time in F, ($_\pi 4_s + _\pi 2_s$ cycloaddition) gives E.

b. (i) $$A \rightleftharpoons D$$

The conrotatory counterpart of the six electron electrocyclic reaction of (a)-(ii) produces the cyclodecapentaene with one trans double bond.

A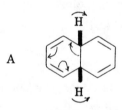

(ii) $$A \rightleftharpoons E$$

The $_\pi 4_s + _\pi 4_s$ cycloaddition reaction is allowed photochemically, so the two butadiene systems of A can cycloadd to give E. An intramolecular exciplex may be formed prior to the cycloaddition.

(iii) $$A \rightleftharpoons F$$

Among the most subtle reactions is the [1,3] sigmatropic shift that interconverts A and F. Starting from F, the product (without redrawing) is A'. The stereochemistry of the 9, 10 bond is appropriate (cis).

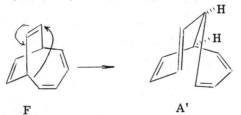

F   A'

Migration with σ symmetry (retention) is allowed photochemically and gives the cis double bond in the product.
The reaction could occur in the ground state (a) with
inversion, but the product double bond would then be
trans and highly strained.

(iv)   B ⇌ E

Similarly, a [1, 3] sigmatropic shift interconverts B
and E. Again the reaction must occur with retention,

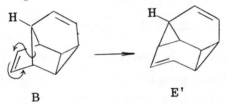

B   E'

since inversion of the tertiary center would place the
indicated bridgehead hydrogen inside the polycyclic
structure. Although these [1, 3] sigmatropic shifts
[(iii) and (iv)] are unlikely possibilities, they are
formally allowed and must be included.

The author does not guarantee that all the possibilities have been
identified, nor have those listed necessarily been observed.
Reference: E. E. van Tamelen and B. C. T. Pappas, _J. Am._
Chem. _Soc_., 93, 6111 (1971).

6. a. The absence of a kinetic dependence on the concentration of
        TCNE indicates that the rate-determining step occurs prior
        to reaction with TCNE (see problem 6-2). Since no mention
        was made of acid catalysis, the slow step must be the iso-
        merization of the starting material A. The only plausible

reaction of A is an electrocyclic ring opening, which is conrotatory for this four electron system (internal cyclo-additions and sigmatropic shifts are unlikely). The structure (B) that results from this isomerization can receive a mole-

cule of TCNE in a Diels-Alder reaction ($_\pi 4_s + _\pi 2_s$ cycloaddi-tion) to give the adduct C, in which there is a cis relationship between the phenyl rings (conrotation).

b. The mechanism can be summarized as follows.

$$A \underset{k_{-1}}{\overset{k_1}{\rightleftharpoons}} B \overset{k_2}{\underset{TCNE}{\longrightarrow}} C$$

Because of the zeroth order dependence on [TCNE], the first step must be rate determining ($\underline{k}_2[TCNE] > \underline{k}_{-1}$). The overall rate is given by

$$\frac{-d[A]}{dt} = \frac{k_1 k_2 [A][TCNE]}{\underline{k}_{-1} + \underline{k}_2[TCNE]}$$

$$= \underline{k}_1[A],$$

since the first step appears to be rate determining. Further-more, it was stated that A is thermally stable. Since isomer-ization to B must indeed occur at the reaction temperature, this observation means that the equilibrium constant, [B]/[A] or $\underline{k}_1/\underline{k}_{-1}$, is well to the side of A, i.e., $\underline{k}_{-1} > \underline{k}_1$. Since formation of the product C occurs in preference to reversion of the intermediate B to the starting material, i.e., the first step is rate limiting, then $\underline{k}_2[TCNE] > \underline{k}_{-1}$. These deductions give the relationship of the three rate constants as $\underline{k}_2[TCNE] > \underline{k}_{-1} > \underline{k}_1$.

Reference: R. Huisgen and H. Seidl, Tetrahedron Lett., 3381 (1964).

7. a. Formation of A is a $_\sigma 2_s + _\sigma 2_s + _\pi 2_s$ cycloaddition reaction. In the alkene a π bond is broken; in quadricyclane two formal σ bonds are broken. Formation of B is a $_\sigma 2_s + _\pi 2_s$ reaction. One σ (quadricyclane) and one π (alkene) bonds are broken. The reaction is demonstrably supra-supra, since the cis disposition of the nitrile groups and of the C–H bonds on quadricyclane is retained.

b. The $_\sigma 2_s + _\pi 2_s$ reaction is forbidden, as can be easily seen by application of the Evans-Dewar method (see problem 11-1). The 2 + 2 + 2 reaction, however, is favorable when addition is suprafacial to all the elements. Consequently, formation of A is preferred.

c.

The same two products are formed, but by different transition states. Formation of A is now a $_\pi 2_s + _\pi 2_s$ reaction, which is forbidden by orbital symmetry. The $_\pi 2_s + _\pi 2_s + _\pi 2_s$ reaction that gives B, on the other hand, utilizes six electrons in a suprafacial fashion and therefore is allowed. Thus quadricyclane can be used to produce A, and norbornadiene can be used to produce B. In each case, the $2_s + 2_s + 2_s$ reaction occurs in preference to the $2_s + 2_s$ reaction.

References: A. T. Blomquist and Y.C. Meinwald, J. Am.

Chem. Soc., 81, 667 (1959); C. D. Smith, ibid., 88, 4273 (1966).

8. Compound A reacts by a retro-Diels-Alder reaction, followed by a [1,3] sigmatropic shift. Compound B first enolizes, and the

bishomotriene unit containing the enol moiety undergoes a homo-Cope rearrangement, followed by a retro-Diels-Alder reaction and aromatization via a [1,5] sigmatropic hydrogen shift. The

retro-Diels-Alder reaction of A probably relieves more strain that the analogous reaction of B, which has an additional carbon atom. Compound A cannot undergo the oxy-homo-Cope reaction because it lacks an enolizable proton. The hydrocarbon homo-Cope in A and B is slow at these temperatures. The electron-donating hydroxyl group reduces the activation energy for rearrangement in the alternative pathway in B.

Reference: T. Miyashi, H. Kawamoto, and T. Mukai, Tetrahedron Lett., 4623 (1977).

9. a. The shortest path (other than the extremely unlikely simultaneous rupture of four bonds) is a reverse Diels-Alder reaction ($_\sigma 4_s + _\sigma 2_s$) to give the intermediate X, followed by a $_\sigma 2_s + _\sigma 2_s$ cycloreversion of X to give the two products. Other intermediates may intervene between X and the products, but none are demanded by the data. The

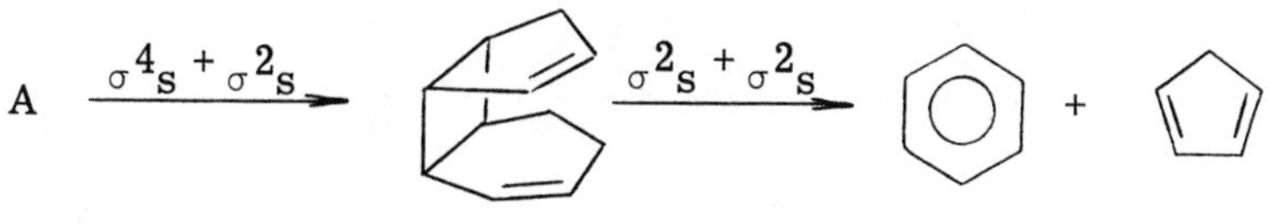

A $\xrightarrow{_\sigma 4_s + _\sigma 2_s}$ $\xrightarrow{_\sigma 2_s + _\sigma 2_s}$ +

X

structure of X is suggested by its resemblance to the trapping product B. The Diels-Alder reaction ($_\pi 4_s + _\pi 2_s$) of X with maleic anhydride gives the product D, which must rearrange via an ene reaction ($_\sigma 2_s + _\pi 2_s + _\pi 2_s$) to B.

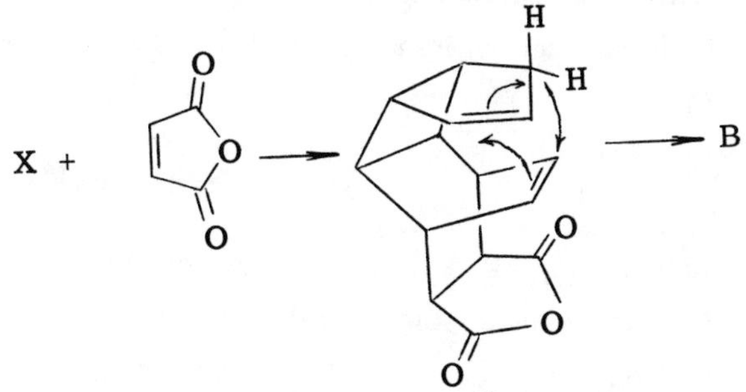

D

b. In the most general case, the formation of the intermediate can be assumed to be reversible. The existence of D, which

$$A \underset{k_{-1}}{\overset{k_1}{\rightleftharpoons}} X \xrightarrow{k_2} C$$
$$M \downarrow k_3$$
$$B$$

precedes B in the trapping step (see (a)), can be ignored in the kinetic scheme. The overall rate must include both k_2 and k_3 and is best expressed simply as loss of A.

$$\frac{d[A]}{dt} = k_1[A] - k_{-1}[X]$$

The steady state in $[X]$ gives

$$k_1[A] = k_{-1}[X] + k_2[X] + k_3[X][M].$$

Thus
$$-\frac{d[A]}{dt} = k_2[X] + k_3[X][M]$$

$$= [X](k_2 + k_3[M])$$

$$= \frac{k_1[A](k_2 + k_3[M])}{k_{-1} + k_2 + k_3[M]}$$

If the first step is rate determining $(k_2, k_3[M] \gg k_{-1})$, the rate expression becomes

$$-\frac{d[A]}{dt} = k_1[A],$$

and the overall rate does not depend on the concentration of maleic anhydride. On the other hand, if conversion of X to C is rate determining $(k_2, k_3[M] \ll k_{-1})$, the rate expression is

$$-\frac{d[A]}{dt} = K_1(k_2 + k_3[M])[A],$$

in which $K_1 = k_1/k_{-1}$ (equilibrium constant for the first step). In this situation, the overall rate increases with the concentration of maleic anhydride. The authors of this study did not comment on the possibility of maleic anhydride dependence on the rate.

References: W. Mauer and W. Grimme, Tetrahedron Lett., 1835 (1976); J. M. Harris and C. C. Wamser, "Fundamentals of Organic Reaction Mechanisms," Wiley, New York, NY, 1976, p 123.

10. a. (i) A disrotatory, six π electron electrocyclic reaction gives a bicyclo[4.2.0]octatriene, which undergoes a self-Diels-Alder reaction. The reverse of this reaction can produce either the original or a new COT.

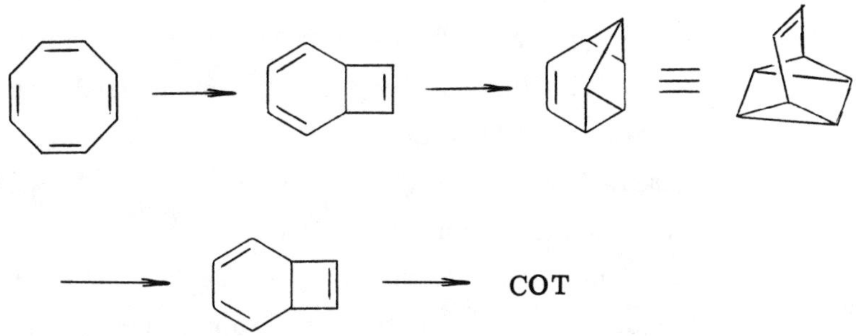

(ii)

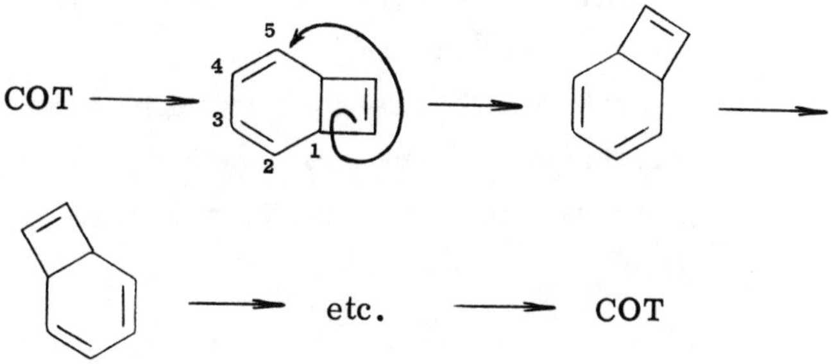

The bicyclic intermediate can alternatively automerize through a [1, 5] sigmatropic shift before returning to COT.

(iii)

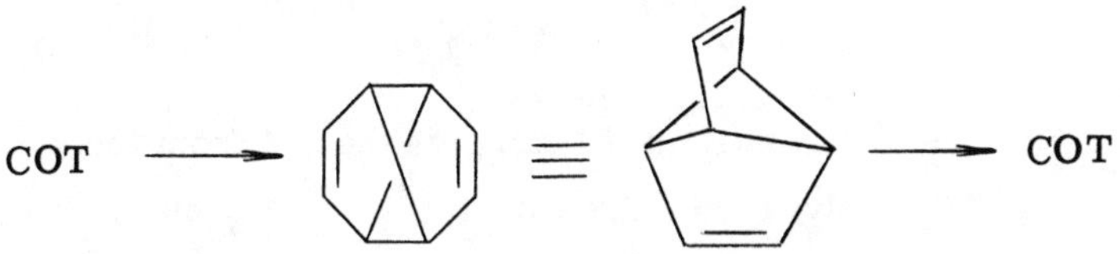

A $_\pi 2 + _\pi 2$ cycloaddition of two opposing double bonds (presumably a stepwise reaction via a bicyclo[3.3.0]-octadienediyl radical) is known to occur in benzo derivatives. Reversal of the reaction with opening of alternative but equivalent bonds in the four-membered ring completes the automerization.

(iv)

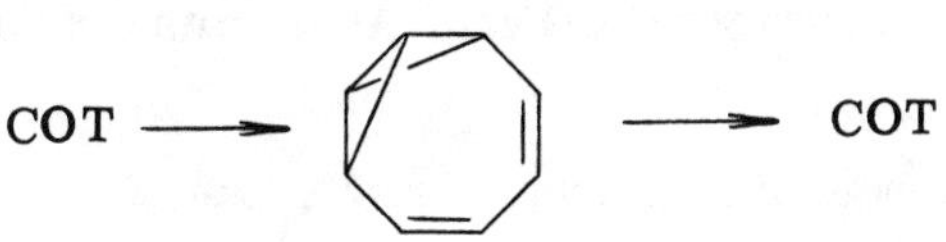

A similar cycloaddition of adjacent double bonds

239

(supra-antara or stepwise) also gives a four-membered ring that can reopen to an automerized form.

There are at least two other possible mechanisms, an antara-antara Cope rearrangement of the bicyclo[4.2.0]octatriene in (i) and (ii) and a $_\pi 2_a + _\pi 2_a + _\pi 2_s$ cycloaddition to give semibullvalene. The latter process does not scramble the carbon skeleton, and the antara-antara Cope is a stereochemically unlikely process when other pathways are available (see problem 11-1).

b. The relocation of the substituent pairs can be determined by carrying the 1,2- and 1,5-dimethyl systems through each mechanism and identifying the conceivable products.

(i)   1,2 $\longrightarrow$ 1,3 + 1,4
        1,5 $\longrightarrow$ 1,4

(ii)   1,2 $\longrightarrow$ 1,4
        1,5 $\longrightarrow$ 1,3

(iii)   1,2 $\longrightarrow$ 1,5
        1,5 $\longrightarrow$ 1,2 + 1,4

(iv)   1,2 $\longrightarrow$ 1,3
        1,5 $\longrightarrow$ 1,4

Either (i) or (ii) can give the 1,4 from the 1,2 and of these only (i) can also give the 1,3. Any of (i), (iii), or (iv) can produce the 1,4 from the 1,5, and of these again only (i) can give the proper products from the 1,2. Therefore it appears that the initial automerization reaction is the electrocyclic/Diels-Alder pathway (i). At higher temperatures, the 1,2 isomer does give the 1,5 and the 1,5 does give the 1,3. Other mechanisms, such as (ii) and (iii), must intervene. The cyclooctatetraene automerization appears to be quite complex, but the electrocyclic/Diels-Alder pathway dominates early in the reaction.

Reference: L. A. Paquette, M. Oku, W. E. Heyd, and R. H. Meisinger, J. Am. Chem. Soc., 96, 5815 (1974).

ALSO SEE PROBLEMS 6-3, 6-8, 7-8, 9-11.

AUTHOR INDEX

241

SUBJECT INDEX

Notes

Notes

Notes